YIELD SIMULATION FOR INTEGRATED CIRCUITS

THE KLUWER INTERNATIONAL SERIES IN ENGINEERING AND COMPUTER SCIENCE

VLSI, COMPUTER ARCHITECTURE AND DIGITAL SIGNAL PROCESSING

Consulting Editor

Jonathan Allen

Other books in the series:

Logic Minimization Algorithms for VLSI Synthesis, R.K. Brayton, G.D. Hachtel, C.T. McMullen, and A.L. Sangiovanni-Vincentelli. ISBN 0-89838-164-9.

Adaptive Filters: Structures, Algorithms, and Applications, M.L. Honig and D.G. Messerschmitt. ISBN 0-89838-163-0.

Computer-Aided Design and VLSI Device Development, K.M. Cham, S.-Y. Oh, D. Chin and J.L. Moll. ISBN 0-89838-204-1.

Introduction to VLSI Silicon Devices: Physics, Technology and Characterization, B. El-Kareh and R.J. Bombard. ISBN 0-89838-210-6.

Latchup in CMOS Technology: The Problem and Its Cure, R.R. Troutman. ISBN 0-89838-215-7.

Digital CMOS Circuit Design, M. Annaratone. ISBN 0-89838-224-6.

The Bounding Approach to VLSI Circuit Simulation, C.A. Zukowski. ISBN 0-89838-176-2.

Multi-Level Simulation for VLSI Design, D.D. Hill, D.R. Coelho. ISBN 0-89838-184-3.

Relaxation Techniques for the Simulation of VLSI Circuits, J. White and A. Sangiovanni-Vincentelli. ISBN 0-89838-186-X.

VLSI CAD Tools and Applications, W. Fichtner and M. Morf, eds. ISBN 0-89838-193-2.

YIELD SIMULATION FOR INTEGRATED CIRCUITS

by

Duncan Moore Henry Walker
Carnegie Mellon University

KLUWER ACADEMIC PUBLISHERS
Boston/Dordrecht/Lancaster

Distributors for North America:
Kluwer Academic Publishers
101 Philip Drive
Assinippi Park
Norwell, MA 02061, USA

Distributors for the UK and Ireland:
Kluwer Academic Publishers
MTP Press Limited
Falcon House, Queen Square
Lancaster LA1 1RN, UNITED KINGDOM

Distributors for all other countries:
Kluwer Academic Publishers Group
Distribution Centre
Post Office Box 322
3300 AH Dordrecht, THE NETHERLANDS

Library of Congress Cataloging-in-Publication Data

Walker, Duncan Moore Henry.
Yield simulation for integrated circuits.

(The Kluwer international series in engineering and computer science. VLSI, computer architecture, and digital signal processing)
Based on author's thesis (Ph. D.).
Bibliography: p.
Includes index.
1. Integrated circuits—Very large scale integration—Design and construction—Mathematical models. 2. Integrated circuits—Very large scale integration—Design and construction—Data processing. 3. Integrated circuits—Very large scale integration—Defects—Mathematical models. 4. Integrated circuits—Very large scale integration—Defects—Data processing. 5. Monte Carlo method. I. Title. II. Series.
TK7874.W34 1987 621.381'73 87-17302
ISBN 0-89838-244-0

Printed in the United States of America

Table of Contents

Preface v

1. Introduction 1

1.1. The VLASIC Yield Simulator 3

1.2. Outline 6

2. Background 9

2.1. Boltzmann, Binomial, and Poisson Statistics 10

2.2. Modified Poisson Statistics 12

3. Defect Models 19

3.1. Process Description 19

3.2. Defect Types 21

3.2.1. Extra Material Defects 22

3.2.1.1. Shorts 23

3.2.1.2. New Devices 24

3.2.1.3. Opens 25

3.2.2. Missing Material Defects 26

3.2.3. Oxide Pinhole Defects 29

3.2.4. Junction Leakage Defects 30

3.3. Model Assumptions 31

3.3.1. Excluded Defect Types 31

3.3.2. Excluded Circuit Faults 32

3.4. Approximations 35

3.4.1. Circles as Octagons 35

3.4.2. Defect Independence 36

4. Defect Statistics 37

4.1. Defect Size Distribution 37

4.2. Defect Spatial Distribution 41

4.2.1. Lot and Wafer Distribution 42

4.2.2. Defect Radial Distribution 45

4.3. Composite Defect Distribution 47

5. Fault Analysis 51

5.1. Preprocessing 51

5.2. Local Circuit Extraction 52

5.3. Layer Combination Analysis 54

5.4. Type-Driven Analysis 56

5.4.1. Shorts 60
5.4.2. Opens 61
5.4.3. New Vias 71
5.4.4. Open Devices 71
5.4.5. Shorted Devices 73
5.4.6. New Gate Device 75
5.4.7. New Active Device 79
5.5. Fault Combination and Filtering 83
6. VLASIC Implementation **87**
6.1. Preprocessing and Data Input 88
6.2. Process Tables 90
6.3. Random Number Generators 91
6.4. Control 92
6.5. Fault Analysis 92
6.6. Fault Combination and Filtering 92
6.7. Fault Summary 94
6.8. Polygon Package 94
6.8.1. Polygon Representation 95
6.8.2. Basic Polygon Operations 96
6.8.3. Polygon Package Algorithms 99
6.8.4. Special Functions 102
6.8.5. Polygon Package Performance 103
6.9. VLASIC Examples 105
6.9.1. Dynamic RAM Cell Example 105
6.9.2. Static RAM Cell Example 109
6.10. VLASIC Performance 111
6.10.1. Random Number Generator Performance 115
6.11. Simulation Accuracy 117
6.11.1. Confidence Intervals for Category Counts 118
6.11.1.1. Approximations to the Binomial Distribution 118
6.11.2. Confidence Intervals for Category Probabilities 120
6.11.2.1. Approximations to the Beta Distribution 121
6.11.3. Halting Criteria 123
6.11.3.1. Halting Criteria for Category Counts 124
6.11.3.2. Halting Criteria for Category Probabilities 125
6.11.4. Random Number Generator Accuracy 125
6.12. Sensitivity Analysis 127
6.12.1. Statistical Model Sensitivity 127

6.12.2. Parameter Sensitivity 129
6.12.3. Fault Detection Sensitivity 129
7. Redundancy Analysis System 131
7.1. Previous Work on Redundancy Analysis 131
7.1.1. Yield Models with Redundancy 132
7.1.2. Yield Simulation with Redundancy 134
7.2. Redundancy Analysis Post-Processor 135
7.2.1. Failure Analysis 136
7.2.1.1. Heuristics 137
7.2.1.2. Cell Failures 139
7.2.2. Spare Swapping Procedure 140
7.3. Redundancy Analysis Examples 141
7.3.1. Comparison to Single DRAM Cell Results 142
7.3.2. Analysis of Heuristics 144
7.3.3. Additional Examples 144
8. Fabrication Data 149
8.1. Process Monitoring 149
8.1.1. Previous Work 151
8.2. Fabrication Data 156
8.3. Model Fitting 157
8.3.1. Defect Size Distribution 157
8.3.2. Radial Distribution 158
8.3.3. Negative Binomial Distribution 160
8.4. Fabrication Data Problems 163
8.4.1. Insufficient Data 164
8.4.2. Missing Data 165
8.4.3. Data Contamination 167
8.5. Economical Process Monitoring 168
8.5.1. Test Analysis Post-Processor 170
9. Conclusions and Current Research 173
9.1. Conclusions 173
9.2. Current Research 175
9.2.1. Faster Monte Carlo Simulation 175
9.2.1.1. Parallel Implementations 176
9.2.1.2. Hierarchical Implementations 177
9.2.1.3. Using More Space 178
9.2.1.4. Fast Simple Fault Analysis 180

9.2.2. Non-Monte Carlo Analysis Techniques 182
9.2.2.1. Critical Area Analysis 182
9.2.2.2. Fault Probabilities and Fault Lists 186
9.2.3. Combining Local and Global Defects 187
References **189**
Index **207**

Preface

In the summer of 1981 I was asked to consider the possibility of manufacturing a 600,000 transistor microprocessor in 1985. It was clear that the technology would only be capable of manufacturing 100,000-200,000 transistor chips with acceptable yields. The control store ROM occupied approximately half of the chip area, so I considered adding spare rows and columns to increase ROM yield. Laser-programmed polysilicon fuses would be used to switch between good and bad circuits. Since only half the chip area would have redundancy, I was concerned that the increase in yield would not outweigh the increased costs of testing and redundancy programming. The fabrication technology did not yet exist, so I was unable to experimentally verify the benefits of redundancy. When the technology did become available, it would be too late in the development schedule to spend time running test chips. The yield analysis had to be done analytically or by simulation.

Analytic yield analysis techniques did not offer sufficient accuracy for dealing with complex structures. The simulation techniques then available were very labor-intensive and seemed more suitable for redundant memories and other very regular structures [Stapper 80]. I wanted a simulator that would allow me to evaluate the yield of arbitrary redundant layouts, hence I termed such a simulator a *layout* or *yield simulator*. Since I was unable to convince anyone to build such a simulator for me, I embarked on the research myself. The redundant microprocessor was never built, but the yield simulation research developed into my Ph.D. thesis [Walker 86a] on which this book is based.

Steve Director patiently pushed my vague ideas into concrete form, led me out of blind alleys, inspired me with his ideas, and pulled this research out of me. I have found our collaboration to be most fruitful and enjoyable.

This research has benefited from discussions with David Black, Ted Equi, Joel Ferguson, Ed Frank, Anoop Gupta, Jean Marie Gutierrez, Leonard Hamey, H. T. Kung, Wojciech Maly, Victor Milenkovic, Sani Nassif, Jim Saxe, Pradeep Sindhu, Bob Sproull and Andrzej Strojwas.

The VLSI group in the CMU Computer Science Department and SRC-CMU Research Center of Excellence for Computer-Aided Design in the Electrical and Computer Engineering Department have provided an environment that made this research possible. This research was funded by the Semiconductor Research Corporation under contracts 82-11-007 and 86-01-068.

Finally, this book is dedicated to my parents, for their love and moral support through the years of research and study.

YIELD SIMULATION FOR INTEGRATED CIRCUITS

Chapter 1
Introduction

Integrated circuit manufacturers find it highly desirable to be able to predict yield loss before a chip is fabricated. The ability to predict yield enables corrective action to be taken before production starts. Corrective action may include changing design rules or process conditions or changing the chip design to add redundancy. The goal of such changes is to maximize the number of good chips per wafer, and thus maximize profits. Waiting for production probe data in order to estimate yield is too costly because it comes much too late in the product cycle.

Process disturbances cause deformations to the ideal integrated circuit (IC). Disturbances cause deformations from the nominal IC structure, such as narrower line width, or extra material. Disturbances that cause local deformations to the IC are called local defects. Disturbances that cause global deformations are called global defects. Local and global defects are the two basic sources of yield loss in the fabrication process. Global defects include layer misregistration and line width variations. Local defects include oxide pinholes and spot defects. It is generally true that global defects primarily affect parametric yield in that they cause variations in speed or power consumption [Maly 81]. The effects of such global defects, which are called parametric defects, can be simulated by the FABRICS II statistical process simulator [Maly 82a, Nassif 82, Nassif 84a]. Local defects, on the other hand, primarily affect circuit topology and cause the chip to fail functionally, i.e. they affect functional yield. Therefore local defects are called catastrophic defects. The changes to circuit topology are called circuit faults. We illustrate the relation between defects and circuit performance in

Figure 1-1. Note that we acknowledge in this figure that global defects can affect functional yield as well as parametric yield and local defects can affect parametric yield as well as functional yield. However these dependencies are small and will be ignored for the remainder of the book. We will discuss this point further in Chapter 9.

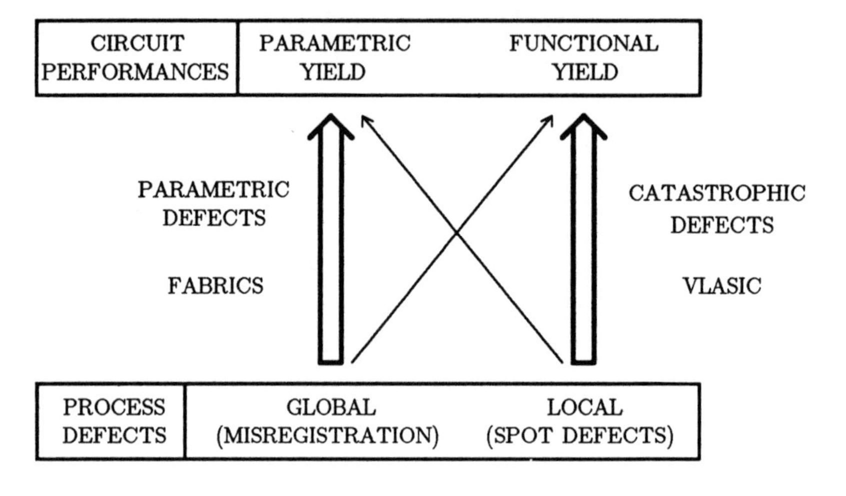

Figure 1-1: Sources of Yield Loss

Because functional yield is determined primarily by catastrophic defects that affect only the layout of a chip, we refer to the method of determining the functional yield as *layout simulation* or *yield simulation*.

The effects of local defects on yield can be determined by generating a population of chip samples that has the same distribution of circuit faults as observed in the fabrication line. This population can be generated using two techniques. In the first technique, the joint probability density function of each circuit fault is determined by analyzing the layout geometry. This probability is combined with the circuit fault statistics observed in the fabrication line to obtain the circuit fault distribution. This approach will be described further in Chapter 9. In this book we determine the circuit fault distribution by using a Monte Carlo yield simulation, where defects are repeatedly generated, placed on the chip layout, and then analyzed to determine what circuit faults, if any, have occurred. In particular, VLASIC, for VLSI LAyout Simulation for Integrated Circuits, is a Monte Carlo yield simulator which has been developed for determining functional yield.

The output of a yield simulator such as VLASIC can be passed to post-processors to predict yield, optimize design rules [Razdan 85], generate test vectors [Maly 84a, Shen 85, Maly 87a], evaluate process sensitivity, etc. We have developed a redundancy analysis post-processor as an example application.

We begin our discussion by describing the major steps required to perform yield simulation, and the overall structure of VLASIC. We then outline the remainder of the book.

1.1. The VLASIC Yield Simulator

The basic steps involved in Monte Carlo yield simulation include the generation and placement of defects on the layout, and the analysis of the modified layout for circuit faults. Several faults may then be combined into a single fault. Uninteresting faults are ignored in a filtering step. The main loop of the Monte Carlo simulation is shown in Figure 1-2. A control loop generates as many chip samples as desired in the simulation. The defect random number generators are used to determine the number and location of defects on the layout with the appropriate statistics. These statistics are obtained from fabrication line measurements.

Once the defects have been placed on the layout, a fault analysis phase determines what, if any, circuit faults have occurred. The resulting faults are passed through a filtering phase that ignores those faults that do not affect the functional yield. The resulting output is a chip sample containing a list of the circuit faults, that have occurred on the chip during the simulated fabrication. The sample is collected in a summary. When simulation is complete, the list of unique chip fault lists and their frequency of occurrence is passed to application post-processors.

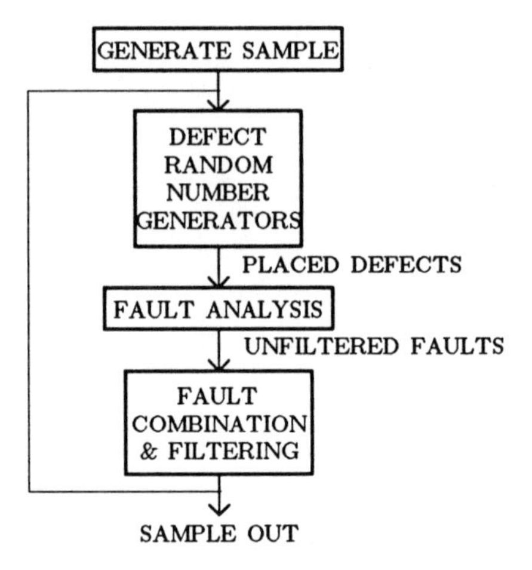

Figure 1-2: Yield Simulator Main Loop

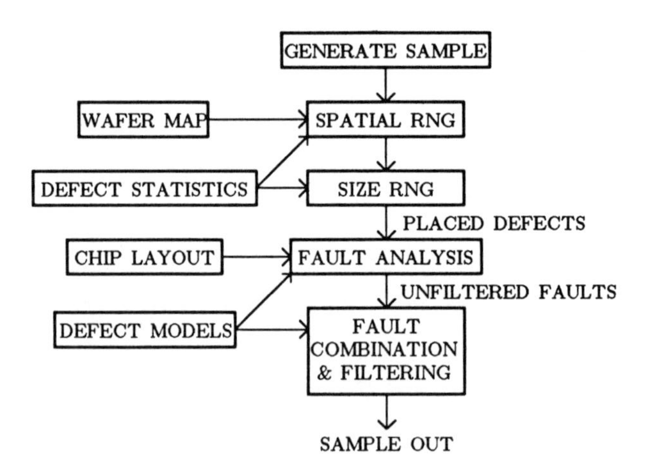

Figure 1-3: VLASIC System Structure

The VLASIC system structure is shown in Figure 1-3. As shown in Figure

1-3, defects have both a size and spatial distribution. A wafer map is used in the process of placing defects within a wafer. The output of the random number generators is a list of defect types, locations within the chip sample, and defect diameters. The defect statistics used by VLASIC are described in detail in Chapter 4.

For each defect, the fault analysis phase calls a series of fault analysis procedures to examine the layout geometry in the defect neighborhood, to determine if any circuit faults have occurred. A separate fault analysis procedure is used for each circuit fault type, such as short or open circuit. The fault analysis procedures manipulate layout geometry using general-purpose polygon operations, so as to be able to handle arbitrary layouts. The result of the fault analysis is a list of the unfiltered circuit faults caused by the defects. The fault analysis procedures are described in detail in Chapter 5.

The unfiltered circuit faults pass through a filtering and combination phase. Those faults that do not cause a DC change to the circuit topology are ignored. An example of such a fault is a new transistor whose terminals are floating. Some faults are combined together into a composite fault. An example of such a situation is when an extra piece of polysilicon breaks an active wire, causing both a new transistor and open circuit. The open circuit information is used to label the transistor source and drain terminals.

The operation of both the fault analysis and filtering phases is guided by defect models. The models specify a description of the fabrication process as a number of patterned layers. Defects are described as modifications to the chip layout, such as circles of extra or missing material. The models also specify what circuit faults can be caused by each defect type, what layers interact with the defect, and how layers are electrically connected together. These models are described in Chapter 3.

After the filtering phase, the resulting output is a chip sample containing a list of the circuit faults, if any, that have occurred on the chip during the simulated fabrication. Each fault on the list specifies what type of fault it was (e.g. short, open), the size, location, and type of the defect causing the fault, the location of the fault in the circuit graph, and how many of that fault occurred. The chip sample fault list is summarized, and then passed to post-processors.

To illustrate the use of VLASIC, we have implemented one application post-processor, a redundancy analysis system, described in Chapter 7. This system takes the unique chip fault lists generated by VLASIC and determines whether the available redundancy can correct the faults.

Our yield simulation research would not be particularly useful if it did not model real fabrication processes. Therefore we have used a set of measurements taken from an operating fabrication line and used them to fit our defect statistical models. In order to simulate a fabrication line, it is necessary to tune the random number generators so that the fault combinations observed on simulated chip samples have the same statistics as those observed in the fabrication line.

1.2. Outline

The outline of the book is as follows. We first describe background research. Next we describe the major components of the VLASIC yield simulator, and then describe its implementation. We then describe a redundancy analysis post-processor. Next we describe how statistical models are fit to a set of fabrication data. Finally we draw conclusions and discuss current research. The major points of the chapters are as follows:

2. *Background* - We describe previous yield modeling work that sets the stage for our research. We describe the limitations of this previous work, and why it does not model reality.

3. *Defect Models* - We describe how we specify a fabrication process. We then describe previous work in determining what types of defects are common in manufacturing. We discuss our techniques for modeling these defects geometrically, and what circuit faults are caused by the defects. We describe qualitatively how defects modify the layout geometry. The details of detecting

the geometry modifications are left to Chapter 5.

4. *Defect Statistics* - We describe previous work on defect statistics that is used in our research. We then describe the composite random number generators used to place defects on the chip layout.

5. *Fault Analysis* - We describe several abortive fault analysis techniques, and why they did not seem promising. We then describe how we use a set of fault analysis procedures using polygon operations to detect circuit faults. We then describe how the faults detected by these procedures are combined and filtered.

6. *VLASIC Implementation* - We describe the structure of the VLASIC implementation, and describe each module, particularly the polygon package. We also describe previous work on polygon packages. We then describe example VLASIC simulations, discuss VLASIC performance, accuracy, and sensitivity.

7. *Redundancy Analysis System* - We describe previous work on yield analysis for redundant circuits. We then describe our redundancy analysis system, give examples of its use, and discuss why our approach is an improvement over previous efforts.

8. *Fabrication Data* - We describe previous work in process monitoring. We then describe the fabrication data available to us. We use this data to select the defect distributions described in Chapter 4. We then discuss the problems with the data, and how better process monitors and analysis tools would make this task easier and more economical.

9. *Conclusions and Current Research* - We discuss the results of this research. We discuss current research directions in improving simulator performance, using new simulation techniques, and making the simulator more accurate by incorporating global as well as local process faults.

Chapter 2
Background

This chapter reviews previous work in the area of yield simulation for the purpose of motivating the research we have undertaken. Most of the work to date has been in terms of generating analytic formulas that at best predict the number of circuit faults on a chip, are limited in that these formulas are only valid for a typical layout [Bertram 83, Glaser 77, Stapper 83a]. More accurate analyses have been done using a combination of analytic and Monte Carlo techniques [Stapper 80, Maly 85a]. Most Monte Carlo analysis has been oriented towards predicting yield for circuits with redundancy [Kung 84, York 85] or design rule optimization [Razdan 85].

Early yield modeling work considered transistors as black boxes with some yield as specified using Boltzmann, binomial, or Poisson statistics. The yield associated with each of these transistors was assumed to be independent of the other transistors [Wallmark 60]. Later actual defects that cause yield loss were taken into account [Hofstein 63]. These models were later refined to more accurately reflect the observed statistics in the manufacturing line [Stapper 83a].

The goals of this early research were much simpler than those of VLASIC. The primary goal was to determine the yield as a function of chip or circuit area. A few researchers also tried to predict the number of circuit faults on a chip. Consequently research efforts focused on determining defect statistics. These statistics were gradually refined until they approached the statistics used in Chapter 4.

The details of the chip layout were not used in yield analysis. Most analyses assumed that the density of circuit faults was proportional to the defect density [Stapper 83a] and so often used the term "defect" to mean a circuit fault. Consequently very simple defect models were adequate, if required at all. There was no need for the complex defect models described in Chapter 3. Since the yield analysis was only in terms of yield or number of faults, there was also no need for fault analysis as described in Chapter 5.

2.1. Boltzmann, Binomial, and Poisson Statistics

A very simple model of IC yield was developed by Wallmark under the assumption that the yield obeyed Maxwell-Boltzmann statistics [Wallmark 60]. He assumed that S out of every 100 transistors were bad, so that the yield of each transistor is

$$Y_1 = 1 - S/100. \tag{2-1}$$

While Wallmark noted that defects are probably correlated, his model assumed no correlation. Thus the yield of N transistors is

$$Y_N = Y_1^{\,N} = (1 - S/100)^N. \tag{2-2}$$

We can view this model as a worst-case yield model.

For the case where $S/100 \ll 1$ and $N \gg 1$[1], Equation (2-2) can be approximated by Poisson statistics and the yield equation becomes

$$Y_N = e^{-SN/100}. \tag{2-3}$$

Hofstein and Heiman observed that gate oxide pinholes were the dominant failure mechanism for silicon MOSFETs, and that these pinholes were randomly distributed across the wafer [Hofstein 63]. Assuming that defects are much smaller than transistor gates, pinholes can be modeled as points, and the number of pinholes in a gate electrode will obey a Poisson distribution. If the defect density is D and the gate oxide area per transistor is a, then the transistor yield becomes

$$Y_1 = e^{-Da}. \tag{2-4}$$

---

[1] An equivalent condition is that $S/100 \rightarrow 0$ when $N \gg 1$ [Gupta 70].

Hofstein and Heiman assumed that gate oxide pinholes were statistically independent, so the yield for N transistors in a circuit is

$$Y_N = Y_1^{\,N} = (e^{-Da})^N = e^{-D(Na)} = e^{-DA}, \tag{2-5}$$

where A is the total gate oxide area of the circuit. The area A where a defect must land to cause a circuit fault is usually called the *critical area*[1] [Stapper 76].

Since the yield Y_N is the probability that a circuit works, the number of working circuits on a wafer is given by the binomial distribution. For this reason, uniformly placed defects are sometimes said to obey binomial statistics [Warner 81, Stapper 81, Stapper 82a]. Dingwall used the same calculation for wafer-scale integration [Dingwall 68].

While the use of Boltzmann and Poisson statistics to characterize defect densities was eventually shown to be too pessimistic for larger chips, due to wafer to wafer and chip to chip variations in defect density [Moore 70], they are still used in many studies [Bernard 78, Borisov 78, Schuster 78, Kovchavtsev 79, Lashevskii 79, Egawa 80, Kitano 80, Posa 81, Sud 81, Mano 82, Saito 82, Uchida 82, Buehler 83, Fujii 83, Mallory 83, Kung 84, Moore 84a, Moore 84b, Sakurai 84, Ueoka 84] [Raffel 85]. This continued use of pessimistic statistics has been due to mathematical convenience and for lack of fabrication line data.

The next phase in developing a yield model came about by recognizing that in fact yield loss may be due to many different defect types rather than one dominant cause. In some processes, the yield loss due to any single defect type is small. All defect distributions are equally good models for the data when the yield is very high [Murphy 64], so it has been suggested that Poisson statistics be used for convenience [Gutierrez 84]. The yield is

$$Y = \prod_i e^{-D_i A_i}, \tag{2-6}$$

[1] The critical area has also been called active area by Hu [Hu 79], susceptible area by Murphy [Murphy 64], and vulnerable area by Walker and Director [Walker 83].

where D_i and A_i are the defect density and critical area respectively for defect type i. Saito and Arai used Equation (2-6) successfully for modeling the yield of MOS test structures and RAMs [Saito 82]. In processes where the yield loss due to each defect type is substantial, Poisson statistics will introduce significant errors into the yield calculation.

2.2. Modified Poisson Statistics

Murphy was the first to note that Poisson statistics resulted in unduly pessimistic yield predictions [Murphy 64]. Poisson statistics are valid when defects are placed uniformly on wafers, but in fact defects tend to cluster, particularly towards the edge of the wafer [Yanagawa 69, Muehldorf 75]. Such clustering leads to yields substantially higher than those predicted using Poisson statistics [Petritz 67]. Murphy observed however, that within small regions with critical area A, the defect density D is constant, so the yield of a circuit in this region obeys Poisson statistics and is given by Equation (2-5). Murphy assumed that D was constant within a chip, and varied from chip to chip with some defect probability density function (PDF) $f(D)$. Thus the overall chip yield is

$$Y = \int_0^\infty e^{-DA} f(D) dD , \tag{2-7}$$

where

$$\int_0^\infty f(D) dD = 1. \tag{2-8}$$

Murphy suggested that $f(D)$ had the form shown by the bold smooth curve in Figure 2-1. For comparison, a δ-function, triangular, and rectangular PDF are shown as curves f_1, f_2, and f_3. The resulting yield from Equation (2-7) for each of these distributions is

$$Y_1 = e^{-D_0 A} \qquad \text{for } f_1 , \tag{2-9}$$

$$Y_2 = \left(\frac{1 - e^{-D_0 A}}{D_0 A} \right)^2 \qquad \text{for } f_2 , \tag{2-10}$$

$$Y_3 = \frac{1 - e^{-2D_0 A}}{2D_0 A} \qquad \text{for } f_3 , \tag{2-11}$$

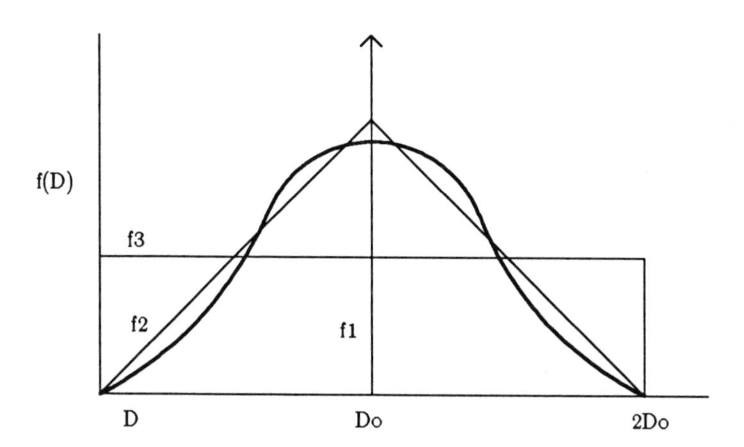

Figure 2-1: Defect Probability Density Functions

where D_0 is the mean defect density. These functions are plotted in Figure 2-2.

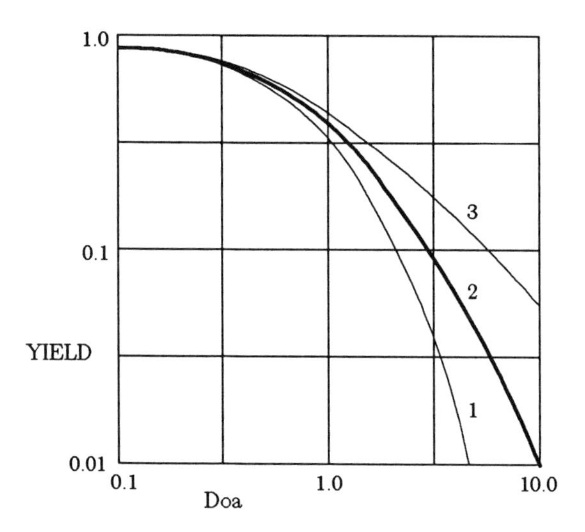

Figure 2-2: Yield Functions

Curve 1 based on Poisson statistics is known to be too pessimistic, and curve 3 predicts yields that are too high [Murphy 64]. Thus curve 2 derived using a triangular PDF best approximates observed yields [Murphy 64].

Ansley showed that even if D is assumed to be constant within a wafer instead of just a chip, Poisson statistics results in yields that are still too pessimistic because D varies from wafer to wafer [Ansley 68]. He confirmed Murphy's result that a triangular PDF best fit the data.

Lawson found that the density D of photoresist pinholes was constant across a wafer, but that D varied normally from wafer to wafer [Lawson 66]. A normal PDF is similar to the bold smooth curve in Figure 2-1. Lawson refined the analysis by using separate critical areas A_i and defect densities D_i for each process step. For each wafer, the yield is then given by Equation (2-6). Yanagawa also found that the defect density between wafers obeyed a normal distribution [Yanagawa 72]. This work confirmed both Murphy's and Ansley's results.

Seeds viewed the fabrication process as a series of events, each of which has an exponential defect PDF

$$f(D) = \frac{e^{-D/D_0}}{D_0}, \tag{2-12}$$

as shown in Figure 2-3 [Seeds 67a, Seeds 67b].

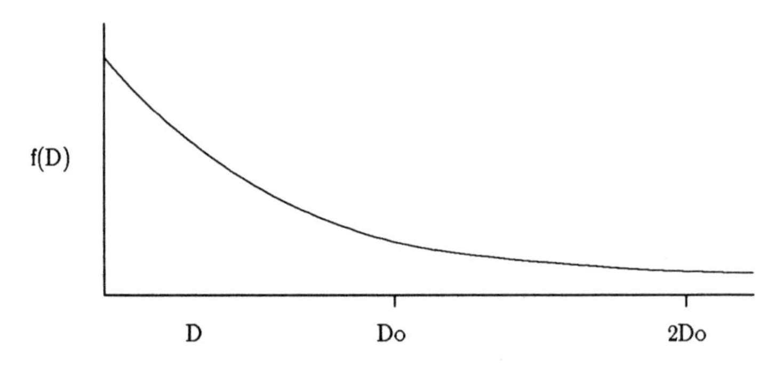

Figure 2-3: Exponential Defect Probability Density Function

Three significant defect types with relative densities 1, 2, and 4 were found to fit the data well. The yield for this model is

$$Y = \frac{1}{(1+A/A_0)(1+A/2A_0)(1+A/4A_0)}, \tag{2-13}$$

where A is the chip area and A_0 is the reference area.

Moore found that integrated circuit yields at Intel could be empirically modeled by the formula

$$Y = e^{-\sqrt{A/A_0}} \tag{2-14}$$

where A_0 is the reference area [Moore 70]. In his paper [Moore 70], Moore included half of a wafer map which indicated the number of defects per chip. "Moore's half slice" was subsequently used as the data set for many yield modeling papers. Equation (2-14) was successfully used through the 1970s at Intel [Warner 81].

Price suggested that Boltzmann statistics were invalid because they assume that defects can be distinguished from one another. He argued that in fact defects were indistinguishable, and therefore Bose-Einstein statistics should be used instead [Price 70]. Assuming Bose-Einstein statistics, the yield is

$$Y = \prod_i \frac{1}{1 + A_i D_i}, \tag{2-15}$$

where A_i is the critical area and D_i is the defect density for defect type i. This equation has the same form as that derived by Seeds in Equation (2-13), and has been used successfully for calculating defect densities from CMOS/SOS test patterns [Jerdonek 78]. Despite this successful use, Price's analysis is incorrect [Murphy 71]. As we will see in Chapter 4, Equation (2-15) often gives the correct results, but for the wrong reasons.

Okabe considered the case of Seeds' exponential density distributions for k process steps, each with the same average defect density D_0 [Okabe 72]. The resulting composite defect PDF is the Erlang distribution PDF

$$f(D) = \frac{(k/D_0)^k}{(k-1)!} e^{-(kD/D_0)} D^{k-1}, \tag{2-16}$$

and the resulting yield is

$$Y = \frac{1}{(1 + D_0 A/k)^k}. \tag{2-17}$$

Okabe assumed that the critical area A is the same for each process step but with different shapes and positions. For the case in which areas exactly overlap, the PDF is

$$f(D)=\frac{(1/D_0)^k}{(k-1)!}e^{-(D/D_0)}D^{k-1}, \tag{2-18}$$

which is the gamma distribution with parameter k, and the corresponding yield is

$$Y=\frac{1}{(1+D_0A)^k}. \tag{2-19}$$

For $k=1$, Equations (2-17) and (2-19) have the same form as Equation (2-15). Equation (2-19) has been used commercially [McMinn 82] and in the laboratory [Borisov 79]. As we will see in Chapter 4, Equations (2-17) and (2-19) are close to those now generally regarded as correct.

Warner noted that different regions on Moore's half slice have radically different yields and suggested considering each region separately, assuming Poisson statistics within each region [Warner 74]. The yield for such a composite model is

$$Y=\sum_i F_i Y_{0,i}{}^{A/A_0}, \tag{2-20}$$

$$\sum_i F_i = 1, \tag{2-21}$$

where F_i is the weight and $Y_{0,i}$ is the yield for unit area A_0 in region i. Since Poisson statistics are assumed in region i, we have

$$Y_{0,i}=e^{-D_{0,i}A_0}, \tag{2-22}$$

where $D_{0,i}$ is the defect density for region i. Equation (2-20) can then be rewritten as

$$Y=\sum_i F_i e^{-D_{0,i}A_0}. \tag{2-23}$$

Warner found that three terms (five free variables) provided an excellent fit to Moore's data. The composite model fits not only the yield, but also the distribution of defects per die. Warner thus became the first person to understand the importance of modeling the fault distribution with a chip as well as the yield [Stapper 83a]. In Chapter 4 we will show how this work was extended to form the foundation of the yield statistics used in VLASIC.

Hu found Equation (2-7) to be incorrect because it used an invalid expansion of the binomial distribution [Hu 79]. Rather than integrating a continuous probability density function, he divided a wafer of area S into regions S_i each with total critical area A and mean defect density D_i. The wafer yield then becomes

$$Y = \sum_i (S_i / S)\, e^{-D_i A}. \tag{2-24}$$

Equation (2-24) has the same form as Equation (2-23). Despite achieving a result that is nearly correct, Hu's approach is invalid since he considers only the yield of a single wafer, rather than many wafers [Stapper 81]. Over many wafers, Equations (2-23) and (2-24) can be approximated as an integration, and the defect densities of many chips approximated as a probability density function, so Equation (2-7) is valid.

Meister showed that for a given average defect density, the Poisson distribution provides a lower bound on the yield [Meister 83]. The upper bound is given by a mixture of Poisson distributions as in Equations (2-23), and by Bose-Einstein statistics as in Equation (2-15). The statistics described in Chapter 4 build on the work of Murphy and Warner to cover the entire yield range in one unified framework.

Chapter 3
Defect Models

This chapter describes the defect models used in our research. These models perform two functions. First, the models provide a geometrical abstraction of local defects. In other words, defects are modeled as geometrical modifications to the specified layout geometry. Second, the models describe how layout geometry combines to form circuit faults. This information can then be used in the fault analysis phase.

We begin by describing the fabrication processes that we are considering and the local defects that can occur in these processes. For each of these defects we then describe the corresponding geometrical model and the type of circuit faults that can result.

3.1. Process Description

We can view the fabrication process as being composed of a repetition of two basic steps: deposition of conducting material and deposition of insulating material. These steps may actually be composed of a number of substeps, such as oxidation, thin film deposition, photolithography, and etching [Sze 83]. Transistor formation is assumed to be self-aligned, occurring where the gate (poly) and active layers intersect. Those active regions not under a gate form diffusion.

In what follows we consider an NMOS process with two metal layers. Extension of what is presented to additional layers of interconnect is straightforward since the same basic process steps are used and, therefore, no new defect types are introduced. Other processes such as CMOS and bipolar involve new process steps, and consequently new defect types. By developing a geometrical abstraction for these defects, and a mapping between defects and circuit faults, these other processes can also be handled by the yield simulation techniques described in this book. We will return to this point in Chapter 9.

The layers and vias for the double-metal NMOS process under consideration are listed in Table 3-1.

```
Layers
    polysilicon
    active
    depletion implant
    buried contact
    first-level metal
    first-level via
    second-level metal
    second-level via
Vias
    metal1-poly
    metal1-active
    poly-active
    metal2-metal1
```

Table 3-1: NMOS Process Characteristics

We assume that second-level metal (metal2) can only connect to first-level metal (metal1), and that metal1-poly vias do not occur on top of transistors or poly-active vias. Vias are assumed to etch to a depth no greater than the surface of lowest connecting layer, but possibly less if connecting to a higher layer. For example, metal1 contacts that ordinarily connect to poly or active can etch as deep as active, but may stop sooner if blocked by poly. Contacts between first and second-level metal etch no deeper than first-level metal. These process assumptions limit the types of circuit faults that can occur, as discussed in Section 3.3.

Since lithography and etching steps have some finite resolution, lines that have less than some layer-specific minimum width Δw are defined as open circuits. Similarly, if a via hole is less than Δw in width, or the top and bottom connecting layers do not both overlap a via hole by at least Δw, then an open circuit has occurred. These situations are shown in Figure 3-1.

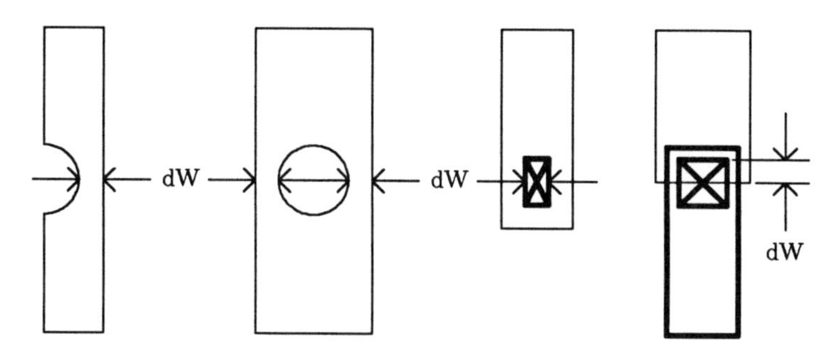

Figure 3-1: Minimum Width Examples

If lines are closer than some layer-specific minimum spacing Δs, then they are assumed to be shorted together. If a line is less than Δs from a contact hole, then the line is assumed to be shorted to the layer at the other end of the contact. These situations are shown in Figure 3-2.

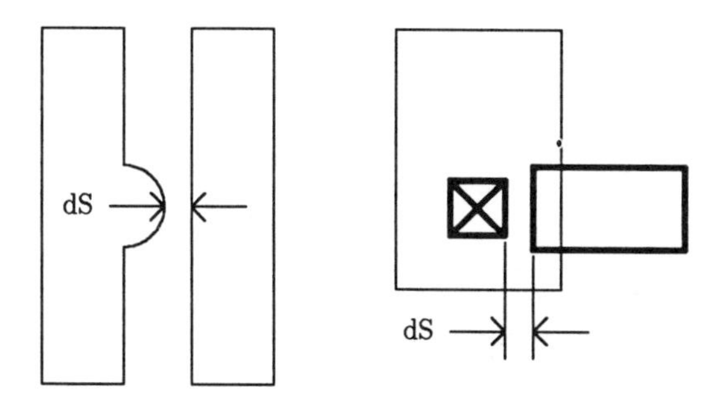

Figure 3-2: Minimum Spacing Examples

Note that these definitions overlap, in that there is a region of uncertainty where a connection may or may not occur. In the discussion that follows, two objects are said to touch if they are closer than Δs and an object is said to be broken if its width at any point is less than Δw.

3.2. Defect Types

Research and manufacturing experience has shown that a few basic types of local defects dominate yield loss in MOS processes: extra and missing material defects, oxide pinholes, and junction leakage [Stapper 76, Stapper

80, Stapper 82b]. Some defects, such as metal hillocks or step coverage breakage, cause local circuit faults, but may be considered global defects since they are caused by global, rather than local, process disturbances. Bipolar processes suffer from other local defects, such as diffusion pipes, which are beyond the scope of this book.

An accurate yield simulation would attempt to simulate how a local process disturbance, such as a crystal defect, causes a local defect, such as a gate oxide pinhole, which may in turn cause a local circuit fault, such as a gate (poly) to substrate short. However the causes of local defects are not well understood. Therefore we will treat local defects as modifications to the layout geometry, without reference to their underlying cause.

3.2.1. Extra Material Defects

Extra and missing material defects are caused primarily by dust particles on the mask or wafer surface, or in processing chemicals. During the photolithography steps, these particles lead to unexposed photoresist areas, or resist pinholes, thus causing unwanted material or unwanted etching of material on a layer [Stapper 80]. For this reason, these defects are sometimes referred to as *photo* or *lithography* defects [Stapper 76]. These defects cause extra and missing polysilicon, active, and metal and may account for 60% or more of the yield loss [Stapper 82c, Mangir 84].

We model an extra material defect as a circular electrically conducting region on the conducting layers, such as poly, active, or metal. Extra material defects can cause short circuits, open circuits, or new devices. A missing via is considered to be a missing material defect, since the via hole is considered to be layer of material. An alternative would be to consider a missing via to be an extra material defect, since the defect causes the via hole to be filled with oxide.

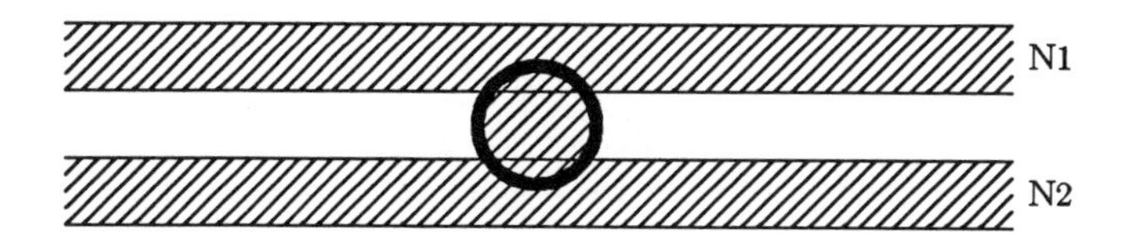

Figure 3-3: N1-N2 Short Due to Extra Poly

3.2.1.1. Shorts

An extra material defect can electrically connect to lines on the same layer, to layers below it through incomplete vias, or to layers above it by blocking a via. An *incomplete via* is one where either the top or bottom layer is missing. These normally occur only in the regions around a buried contact where the buried layer intersects poly or active, but not both, as shown in Figure 3-4. A *blocked via* is one that connects a layer above the extra material defect to a layer below the defect, and the defect partially or completely blocks the connection, causing both the upper and lower layers to short to the defect, or only the upper layer.

An intralayer short is shown in Figure 3-3, with an extra poly defect shorting to two lines. Both an intralayer and an incomplete via short are shown in Figure 3-4 with extra poly shorting to two lines, and to active through a buried contact.

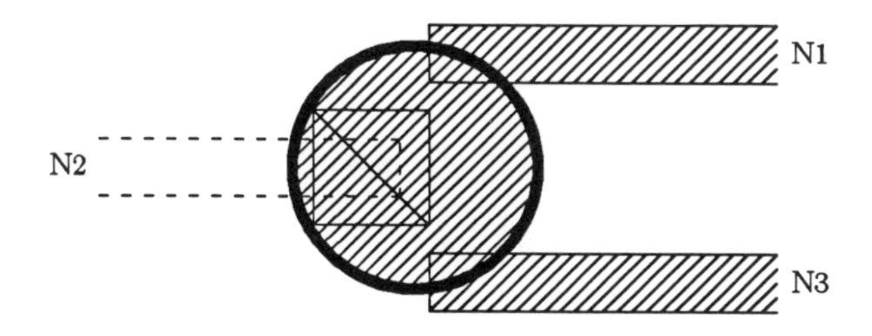

Figure 3-4: N1-N2-N3 Short Due to Extra Poly

An intralayer and partially blocked via short are shown in Figure 3-5 with extra poly connecting to a line and to metal and active by partially blocking a via. Since only DC changes to circuit topology are of interest, a short is only recorded if a defect connects two or more nets with different net

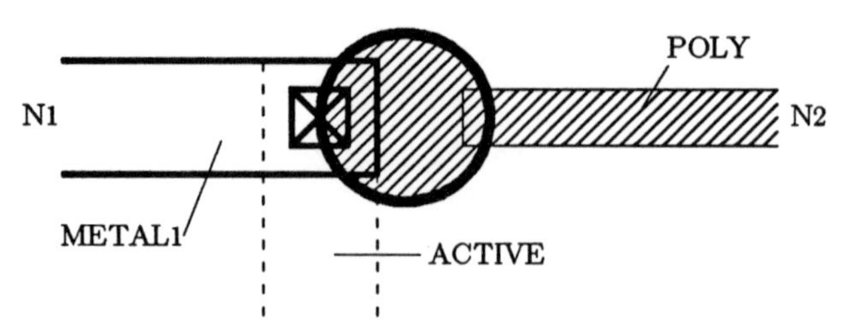

Figure 3-5: N1-N2 Short Due to Extra Poly

numbers. If a short connects two lines that are part of the same net, it can be ignored because it only changes the net resistance.

3.2.1.2. New Devices

The poly and active layers are not independent since their intersection forms a transistor. An extra poly defect that spans an active line inserts a transistor in series with the line, as shown in Figure 3-6.

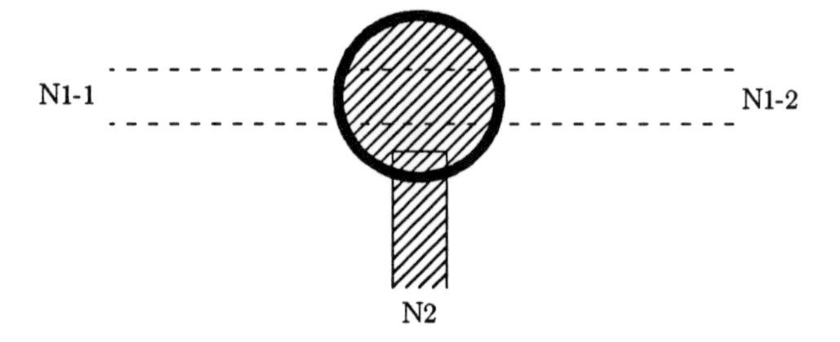

Figure 3-6: New Gate Device

An extra active defect forms a new transistor if it passes under a poly line, as shown in Figure 3-7.

Since we are only considering faults causing a change to the DC circuit topology, in order to be a valid device, a new transistor must have both its source and drain connected. Otherwise it is a capacitor, and is ignored. Similarly, if extra active or poly simply modifies a transistor channel shape, as shown in Figure 3-8, it is ignored.

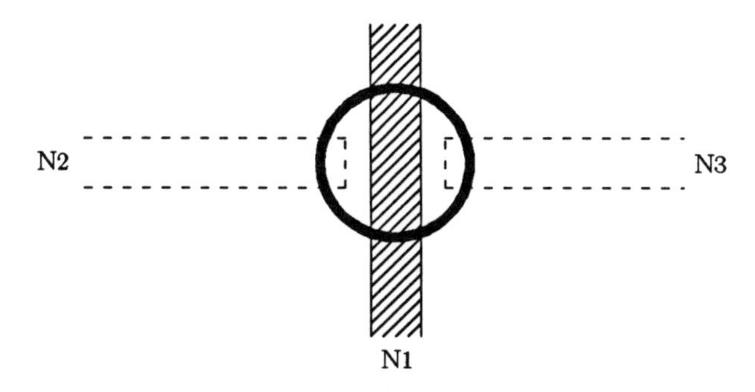

Figure 3-7: New Active Device

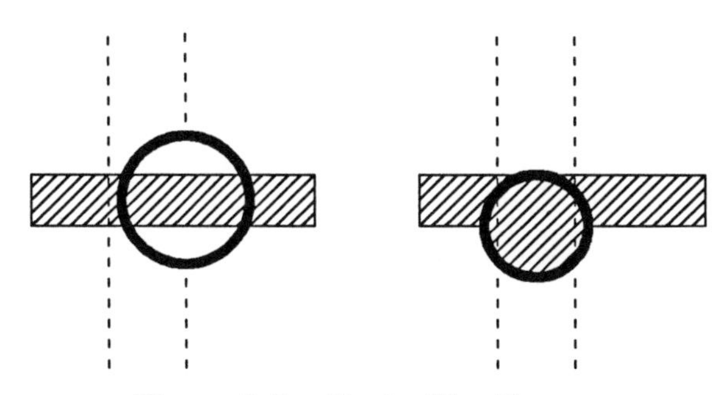

Figure 3-8: Device Size Changes

If the extra poly occurs at a fork in an active net, or the extra active occurs at a fork in a poly net, then a *multi-terminal* device (abbreviated as *multi* [Gupta 81]) can be created as shown in Figure 3-9. Multi-terminal devices have more than two source/drain terminals. In this book, a *terminal* always refers to a transistor source or drain. A transistor gate will always be referred to as a *gate*. The multi-terminal device can also involve some neighboring transistors, in which case the transistors are said to be *converted* to a multi-terminal device as shown in Figure 3-10.

3.2.1.3. Opens

An extra poly defect will cause an open circuit if it breaks an active line in the process of forming a new transistor, as shown in Figure 3-6. If the result is a multi-terminal device, then the net is broken into more than two pieces. Each of these pieces of the broken net is called a *branch*. An extra material

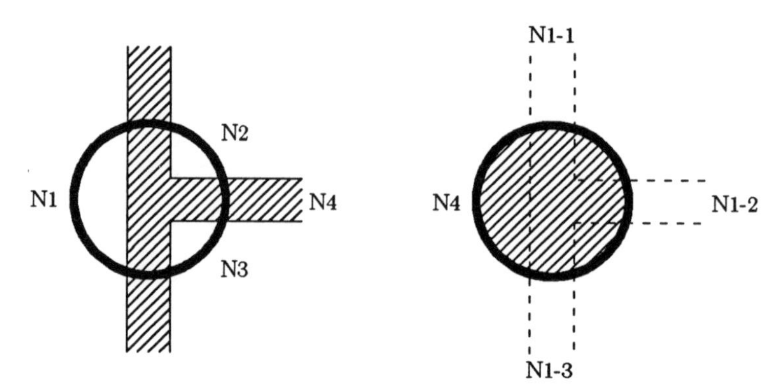

Figure 3-9: New Multi-Source/Drain Devices

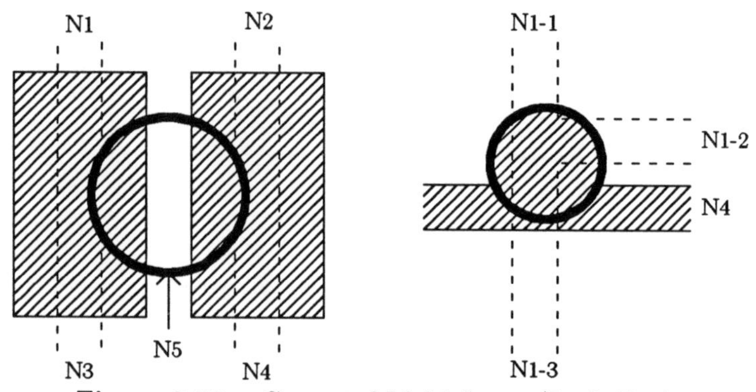

Figure 3-10: Converted Multi-Source/Drain Devices

defect can also cause an open if it completely covers a via, so the lower layer is no longer connected to the upper layer, causing a blocked via. The only case where this can happen in the NMOS process is an extra poly defect blocking a first-level metal to active via, as shown in Figure 3-11.

3.2.2. Missing Material Defects

As mentioned in the previous section, missing material defects are caused primarily during the lithography process by dust particles. We assume that missing material may cause both open lines and open vias. Some authors have treated open vias separately from open lines [Turley 74, Ipri 77, Ipri

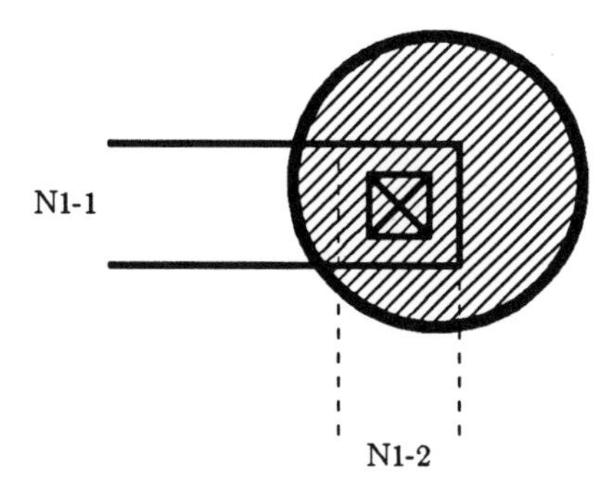

Figure 3-11: Open Circuit Due to Blocked Via

79, Linholm 81]. Some authors have considered open vias to be caused by global process disturbances, such as incomplete etching [Stapper 80].

Open lines can be caused by second metal crossover breakage, which can be significant for some processes, particularly for narrow lines [Turley 74]. This breakage can be considered a missing material defect. As mentioned previously, we consider poor step coverage to be a global defect, and hence do not include it in this research. In addition, modern oxide planarization techniques minimize this problem.

We model a missing material defect as a circular electrically insulating region within a conducting layer, such as poly, active, or metal, and also within a via layer, such as metal2-metal1 or metal1-poly. Missing material defects can cause open circuits, open devices, and shorted devices. A missing material defect can break a line on the same layer if it spans the line, as shown in Figure 3-12. If the defect type is missing material on a via layer, then an open will occur if the defect covers a via, as shown in Figure 3-13. If a missing material defect occurs at a fork in a line, then the net will be broken into several branches. If a missing active defect spans a transistor channel from side to side, then the channel will be missing, causing an open device, as shown in Figure 3-14. If a missing poly defect spans the channel from source to drain, then a shorted device occurs. Due to the self-aligned nature of the process, this region is filled in with diffusion, causing the source and drain to short together, as shown in Figure 3-15.

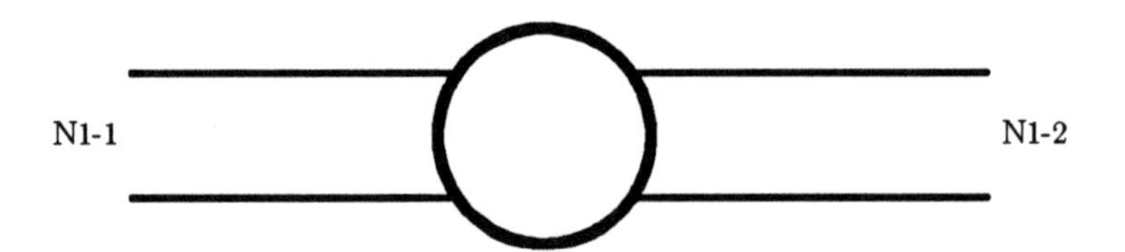

Figure 3-12: N1 Open Due to Missing Metal

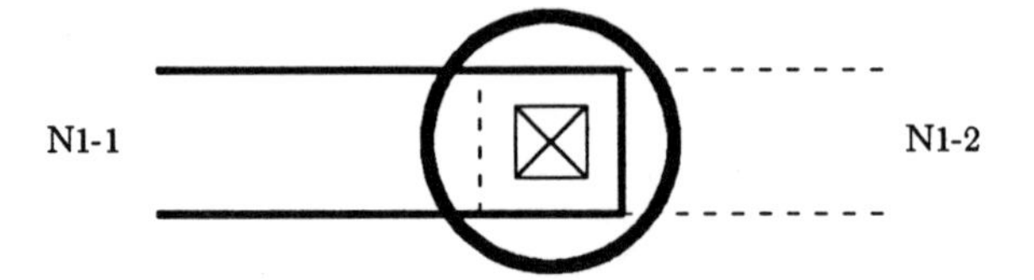

Figure 3-13: N1 Open Due to Missing Via

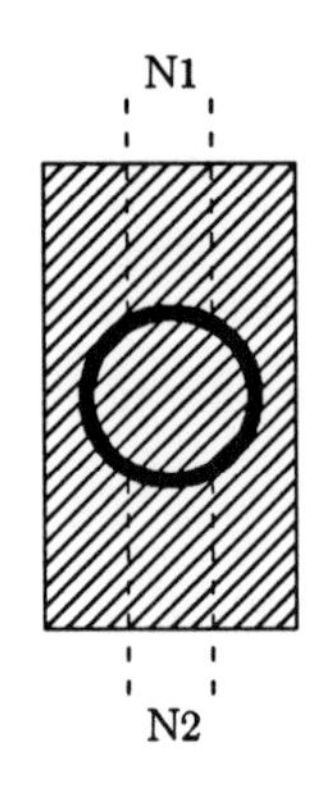

Figure 3-14: Open Device

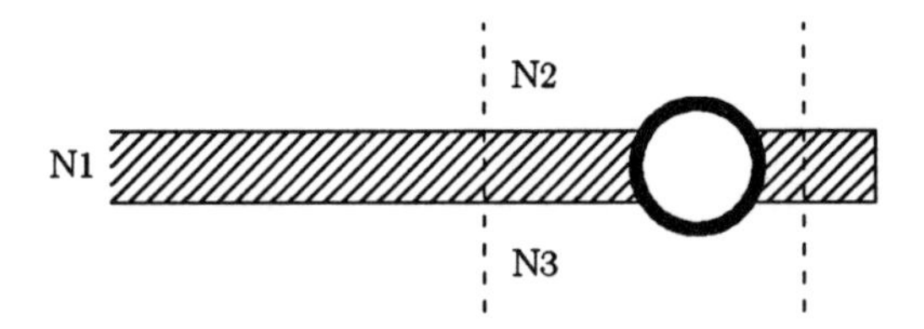

Figure 3-15: Shorted Transistor

3.2.3. Oxide Pinhole Defects

The principle cause of poor yield in gate oxides is the presence of pinholes due to oxygen deficiencies at the Si-SiO_2 interface, tensile stress, surface imperfections, chemical contamination, nitride cracking during field oxidation, and crystal defects [Williams 81, Hofstein 63, Kovchavtsev 79]. The type of substrate dopant does not affect the defect density [Lashevskii 79]. The presence of a buried contact mask sharply increases pinhole densities [Saito 82].

Most research indicates that gate oxide yield is a function of oxide area. However, depending on process conditions, oxide yield is a function of gate to field oxide edge length or source/drain edge length [Lashevskii 79, Saito 82]. We assume that gate oxide pinholes are only a function of area.

A primary cause of oxide pinholes in the first intermediate oxide is resist stripping at the highest parts of the chip (in this case, the resist on top of poly lines) during contact printing [Saito 82]. Projection printing greatly reduces the incidence of these defects, and eliminates the correlation between high points and pinhole location. We will assume that oxide pinholes in intermediate oxides are not affected by chip topography.

Hillocks are sometimes a major cause of first to second metal shorts in large metal overlap areas, so that large parallel plate capacitors have a higher failure rate than overlapping first and second metal lines with a small overlap area. However Turley found the opposite to be true [Turley 74]. Hillocks can be considered as being electrically equivalent to oxide pinholes since they form conducting vias through the oxide. As mentioned previously, we consider hillock formation to be a global defect, and therefore do not include it in this research.

We model oxide pinholes as points where oxide is missing. If these points intersect a region where two conductors overlap, then the conductors are shorted together, as shown in Figure 3-16. As in the case of extra material defects, a short can only occur between two different nets. In the case of a

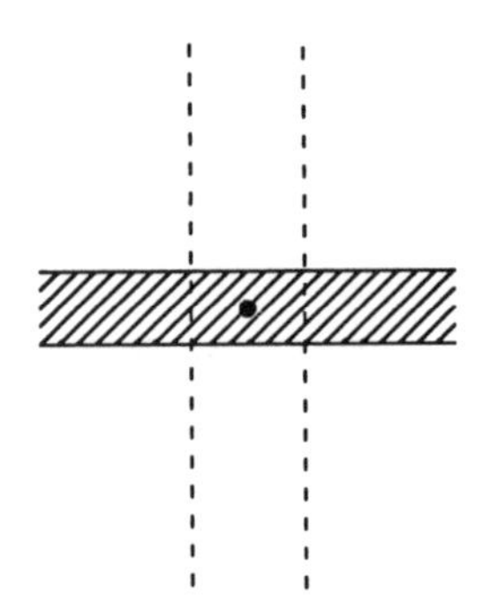

Figure 3-16: Oxide Pinhole Short

gate oxide pinhole, the gate net is shorted to a special channel net *NCHAN*. Since the electrical meaning of a gate (poly) to channel short is not well defined, it is classified as a *new via* rather than a short. An oxide pinhole is only assumed to be able to extend through one deposited oxide layer. Examples of such oxides are the gate oxide, poly-metal1 oxide, and metal1-metal2 oxide. The latter two oxides are usually referred to as the first and second intermediate oxides.

3.2.4. Junction Leakage Defects

Junction leakage defects occur in diffused regions where crystal defects or contamination have occurred at the junction [Jastrzebski 82]. At these points, high fields cause substantial current leakage. Junction leakage defects are a significant source of yield loss, particularly in DRAMs [Stapper 76, Stapper 80, Stapper 82b, Jastrzebski 82]. Since diffused regions have a side wall, junction leakage defects are a function of the diffusion periphery, as well as area. In our research, we ignore this edge-related phenomenon, as we have no data measuring it, nor has any been published in the literature.

Leakage defects are modeled as points where a junction cannot form. If these points intersect a diffusion line, then the line is shorted to the substrate, as shown in Figure 3-17. The substrate net is denoted by the special net *NSUB*.

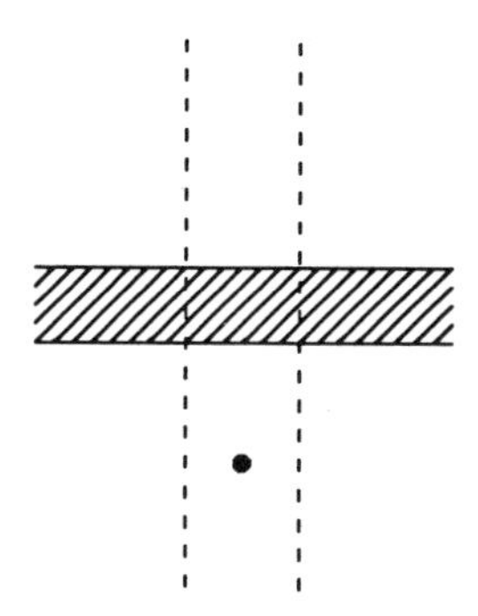

Figure 3-17: Junction Leakage Short

3.3. Model Assumptions

Our defect model is not fully general. We have already restricted ourselves to defects that occur only in NMOS processes. In addition, we make some assumptions that allow us to ignore rare and difficult-to-handle defect types or circuit faults. In most cases, existing processes satisfy these assumptions.

3.3.1. Excluded Defect Types

We do not consider defects in the implant, buried contact, or overglass layers in this book. Excluding implant layer defects means that we can ignore whether a transistor is enhancement or depletion-mode. Buried contact defects do occur in some processes [Saito 82], but have not been a problem in the process described in Chapter 8.

Extra contact cuts are also not considered in this book. Double contact masks can be used to minimize this type of defect.

We do not consider crystal line defects in the substrate that lead to shorts between adjacent diffusion lines. There is little information on this defect type in the literature.

We assume that there is only one active layer and one gate-forming (poly) layer. We assume that all transistors in the original circuit have two source/drain terminals. This assumption implies that capacitors are not allowed. These assumptions are merely a convenience to simplify the fault analysis procedures, rather than a fundamental limitation of the analysis techniques.

3.3.2. Excluded Circuit Faults

When short circuits are reported, we assume that the defect causing the short cannot simultaneously cause an open in any of the shorted nets. This assumption is sometimes invalid for the case in which a blocked via occurs. For example, an extra poly defect can break an active net N1 into two pieces N1-1 and N1-2. At the same time, the defect can block a metal-active contact, shorting the defect to piece N1-3. If the defect also shorts to net N2, then the defect has simultaneously caused net N1 to be broken into pieces N1-1, N1-2, and N1-3, and shorted net N2 to N1-3. This example is shown in Figure 3-18.

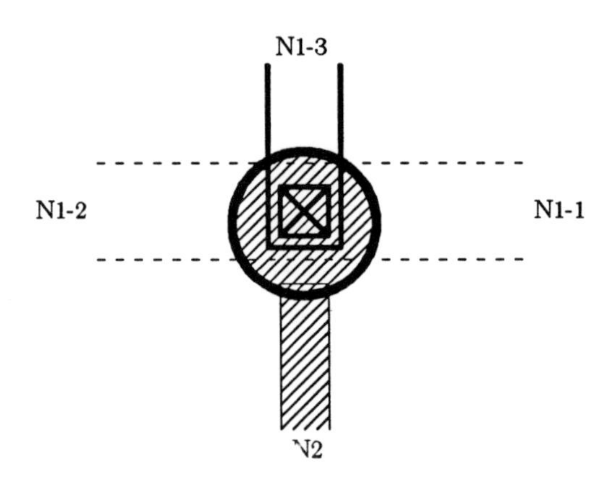

Figure 3-18: Shorted Net and Net Fragment

In order to simplify fault analysis, we consider shorts and opens separately. Consequently the faults that are reported are N1 broken into N1-1, N1-2, and N1-3, and N1 shorted to N2. The correct analysis would report N1-3 shorted to N2. This situation also occurs when a buried contact causes the extra poly defect to be shorted to one branch of the broken active net N1 while at the same being shorted to N2. Since in our example NMOS processes, a blocked via only occurs when extra poly blocks a metal-active via, it can be detected as a special case.

Another situation where a defect simultaneously causes a short and an open is when a missing poly defect converts a metal-poly contact into a metal-active contact. In Figure 3-19, the missing poly causes nets N2 and N3 to be shorted together, breaks N1 into N1-1 and N1-2, and shorts N1-1 to N2 and N3.

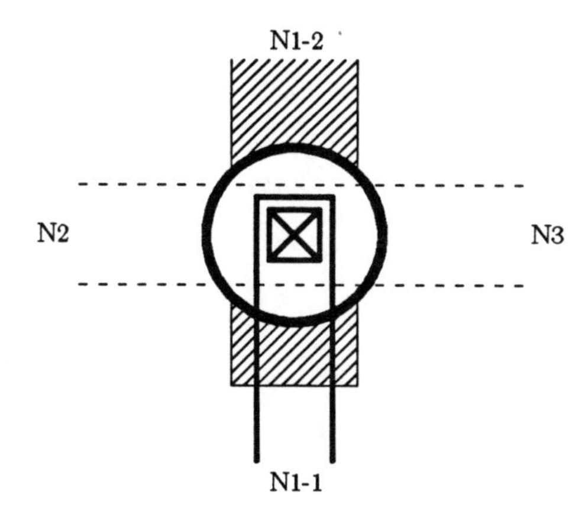

Figure 3-19: Shorted Nets and Net Fault

For poly-gate MOS processes, this situation can only occur if metal-poly contacts are allowed over transistor channels. Most processes do not permit this configuration, so we assume it does not occur.

Figure 3-19 is an example of an *unblocked via*, where a missing material defect removes the material from the bottom of a via, allowing the etching process to continue down to a lower layer. Our definition of the etching process permits a metal-poly via to be unblocked, creating a metal-active via. However a metal1-metal2 via cannot be unblocked since metal2 only connects to metal1. Since we assume that metal-poly vias do not occur on top of transistors, unblocked vias cannot occur in our example NMOS process. We can then make the assumption that missing material defects cannot cause shorts, except for shorted devices.

Both unblocked and blocked vias are cases where a defect causes both an open and short circuit. The only other case where both a short and an open occur is when an extra poly defect breaks an active net, forming a new transistor, as well as shorting several other nets together.

We assume that only extra material defects can add new terminals to existing transistors. However if a metal-poly contact occurs on top of a transistor channel, then a missing poly defect can add a new terminal as shown in Figure 3-20.

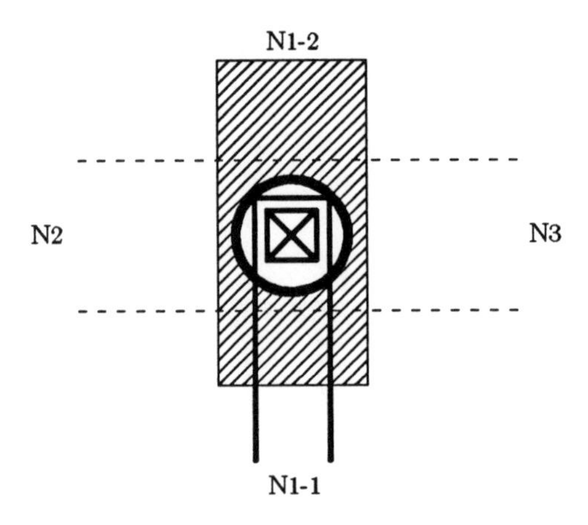

Figure 3-20: New Transistor Terminal

Net N1 is broken into N1-1 and N1-2. The missing poly forms a new diffusion area in the center of the transistor which is connected to N1-1, creating a new terminal. If metal-poly contacts are not allowed on top of transistors, as previously assumed, then missing poly cannot cause new transistor terminals to be created. If missing poly did create a "hole" in the middle of a transistor, it would be left floating. In our transistor analysis, we discard floating source/drain terminals. Therefore this new terminal is ignored, so the transistor terminal count remains unchanged.

3.4. Approximations

The process description and defect model are not employed exactly as described above. For the sake of simplicity or efficiency, some further approximations are required. These approximations, which are described in this section, introduce small errors into the analysis. However, an effort has been made to quantify the errors and so justify the approximations.

3.4.1. Circles as Octagons

Since the fault analysis phase uses polygon operations, the circular material defects must be approximated by polygons. We do not use circles because polygon packages that incorporate arcs as well as line segments, such as described in [Sutherland 78] are very slow [Lang 79]. Hence, circles are inscribed in octagons with its faces oriented at 45 degree angles as shown in Figure 3-21.

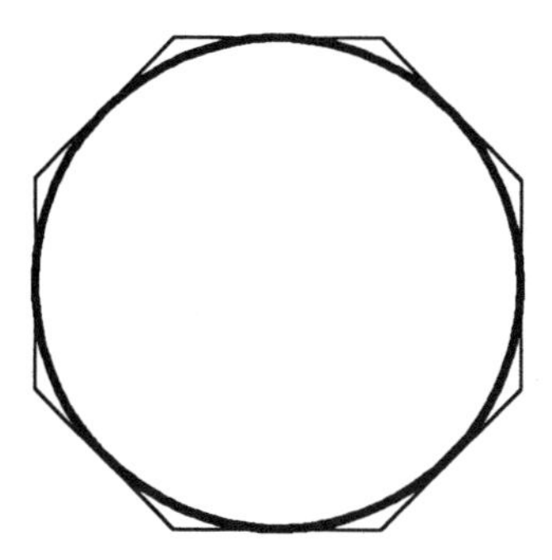

Figure 3-21: Octagonal Approximation of a Circle

The face orientation means that for parallel lines oriented at a multiple of 45 degrees, there is no error in calculating whether a short or an open has occurred. Errors occur only for interactions with objects that are not at a 45 degree angle from the defect center. The maximum diameter of the octagon is 1.082 times that of the circle. The octagon area is 1.055 times greater than that of the circle. The error introduced by the octagon approximation should be, on average, much less than 5.5%. Other researchers have approximated defects as rectangles [Noice 81], but we felt that the error introduced would be unacceptably large.

3.4.2. Defect Independence

Defects are rare enough so that we can assume that no two defects ever interact. It may be impossible in a process for some defects to interact, such as extra and missing material at the same location. However it is possible for a missing material defect to break a line in one location, and for one of the resulting branches to be shorted to some other line. Even in a low yield process, there are rarely more than a few tens of circuit faults on a chip. Given the tens or hundreds of thousands of nets and devices, the probability of two circuit faults affecting the same net or device is small enough to neglect. In addition, faults like short circuits can be considered separately and later merged together. That is, if net N1 is shorted to net N2, and net N2 is shorted to net N3, then these shorts can later be combined into a N1-N2-N3 short.

The only place where the independence assumption may be a problem is in faults that involve global nets, such as power and ground nets. These nets can consume a large fraction of the total net length, and so are involved in a large fraction of the circuit faults. Shorts to the power and ground nets are usually assumed to cause a circuit node to be stuck at 0 or 1, rather than affecting the nets. The obvious exception is shorts between power and ground. However it may be the case that a power or ground net can be open in several places. This situation normally results in a complete chip failure, so the details of the fault are not critical.

Chapter 4
Defect Statistics

In this chapter we describe the statistical distributions used to model how the defects described in Chapter 3 occur on a chip. Local defects can be characterized by a spatial distribution and a size distribution. The spatial distribution describes how defects are distributed across lots, wafers, and chips. The size distribution describes how the defect diameter varies.

In what follows we make the assumption that each defect type may be treated independently. This is a common assumption [Stapper 83a]. Thus defects that occur during one process step do not affect the probability of a defect occurring in a later process step. The data from an actual fabrication process, as described in Chapter 8, substantiates this assumption. Note that this independence assumption can only be made for the local defects considered in this book. Some global defects can clearly affect several layers. For example, inadequate planarization may cause step coverage breakage on several interconnect layers.

4.1. Defect Size Distribution

In Chapter 3 we modeled extra and missing material defects as circles. The diameter of these circles is characterized by a size distribution. Some research assumed that defects had a constant diameter [Lawson 66, Hu 79]. However most research has shown that small defects are more common than large defects [Ipri 77], which is also true of particles from the air or chemicals that are deposited on the wafer. Below some diameter, defect densities fall to zero as defects can no longer be resolved by the process.

Bernard found that on chromium masks, half of the defects were smaller than five microns in diameter [Bernard 78]. The frequency of defects fell to zero above and below four microns in diameter. Some researchers assumed that defect density fell off as $1/x^2$ with increasing defect diameter x [Schuster 78] while others assumed a $1/x$ distribution [Barton 80a]. Dennard found that the frequency of lithography defects falls off as $1/x^3$ with increasing diameter [Stapper 76, Stapper 80, Stapper 82b]. This held true for defects larger than 1.8 microns, and the assumption was made that the density peaked and then fell to zero somewhere below this value.

By employing optical microscope observations of memory chips, Stapper was able to determine the defect size distribution [Stapper 83b]. Very small defects could not be observed, but their frequency was assumed to rise linearly to some peak at diameter x_0. Beyond this value, the frequency was observed to fall off as $1/x^n$ with increasing size, as shown in Figure 4-1.

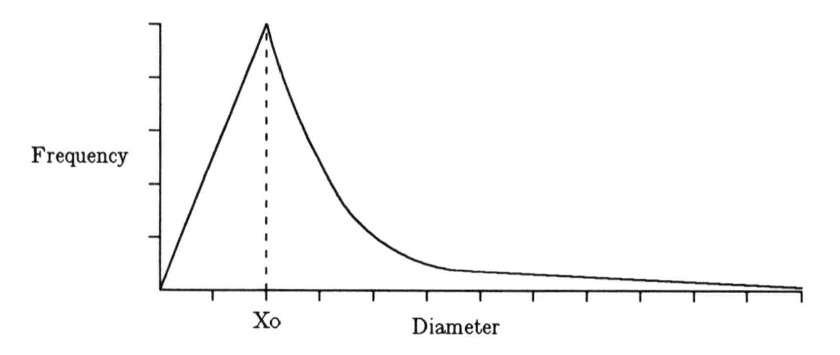

Figure 4-1: Defect Diameter Size Distribution

The normalized distribution function is

$$h(x) = \frac{2(n-1)x}{(n+1){x_0}^2}, \qquad 0 \leq x \leq x_0, \tag{4-1}$$

$$h(x) = \frac{2(n-1){x_0}^{n-1}}{(n+1)x^n}, \qquad 0 \leq x \leq \infty. \tag{4-2}$$

Stapper fit Equations (4-1) and (4-2) to process monitor data [Stapper 84a], with the best fit being $n = 3.02$. This is quite close to Dennard's value of $n = 3$. For $n = 3$, Equations (4-1) and (4-2) become

$$h(x) = \frac{x}{x_0^2}, \qquad 0 \le x \le x_0, \tag{4-3}$$

$$h(x) = \frac{x_0^2}{x^3}, \qquad x_0 \le x \le \infty. \tag{4-4}$$

The actual defect size distribution $D(x)$ is obtained by multiplying $h(x)$ by the mean defect density D.

Ferris-Prabhu considered a more general distribution than Stapper [Ferris-Prabhu 85a, Ferris-Prabhu 85b, Ferris-Prabhu 85c]. The number of defects was assumed to vary as some inverse power p of defect size above diameter x_0, and as some power q below x_0. The size distribution is

$$S(x) = \frac{cx^q}{x_0^{q+1}}, \qquad 0 \le x \le x_0, \; q \ge 0, \tag{4-5}$$

$$S(x) = \frac{cx_0^{p-1}}{x^p}, \qquad x_0 \le x \le x_M, \; p \ge 0, \tag{4-6}$$

$$S(x) = 0, \qquad x \ge x_M, \tag{4-7}$$

$$c = \frac{(q+1)(p-1)}{(q+p) - (q+1)(x_0/x_M)^{p-1}}, \quad p \ne 1, \tag{4-8}$$

$$c = \frac{q+1}{1 + (q+1)\ln(x_M/x_0)}, \qquad p = 1, \tag{4-9}$$

where x_M is the maximum defect size, which is assumed to be much larger than the minimum line width w, but smaller than the die size. The diameter of peak defect frequency x_0 is assumed to occur at a diameter smaller than w. When $q = 1$, $p = 3$, and $x_M = \infty$, Equations (4-5) to (4-9) are the same as Equations (4-3) and (4-4).

Maly suggested that the diameter of defects was a function of both the diameter of defects on the mask and the line width distortion that occurs when transferring the mask pattern to the IC [Maly 84b, Maly 85b, Maly 85c]. Previous work had only attempted to model the combined effects. Measurements showed that there were two defect sources: dust particles from the air, and mechanical contact of the mask with wafers and processing equipment. Therefore the probability density function of defect diameters had several components. The diameter PDF of defects from a single source was modeled by a Rayleigh distribution

$$f_r(x,\alpha) = \frac{x}{\alpha^2} \exp\left(-\frac{x}{2\alpha^2}\right), \qquad x > 0, \tag{4-10}$$

$$f_r(x,\alpha) = 0, \qquad x < 0, \tag{4-11}$$

where α is the distribution parameter. The composite PDF of mask defect diameters was modeled by

$$f(R) = [\beta_1 f_r(R - m_1, \alpha_1) + (1-\beta_1) f_r(R - m_2, \alpha_2)]\beta_2 + (1-\beta_2) f_r(R - m_3, \alpha_3). \tag{4-12}$$

β_1 and β_2 are weights. m_1, m_2, α_1, and α_2 characterize defects caused by dust particles. m_3 and α_3 describe defects caused by mechanical contact. In a modern process, $\beta_2 = 1$ since contact printing is not used. This distribution is convolved with a normal distribution describing the line width distortion to obtain the distribution of defect sizes appearing on the wafer surface. This composite distribution was used to fit the defect size distribution reported in [Stapper 83b].

Chip layouts are typically designed to obey some design rules. For example, all metal lines must have some minimum spacing. Extra metal defects that are significantly smaller than this spacing cannot cause metal lines to short together. In all real processes, the minimum width and spacing rules are substantially greater than the diameter of peak defect frequency. This is so because it is not possible to manufacture parts with widths and spacings at the resolution limits of the photolithography process, where the peak defect frequency occurs. Therefore it is possible to only consider defects larger than size S. If this is done for Equations (4-3) and (4-4), the defect frequency would then be zero if the diameter is less than S and fall off as $1/x^3$ above S with a correspondingly lower defect density, as shown in Figure 4-2.

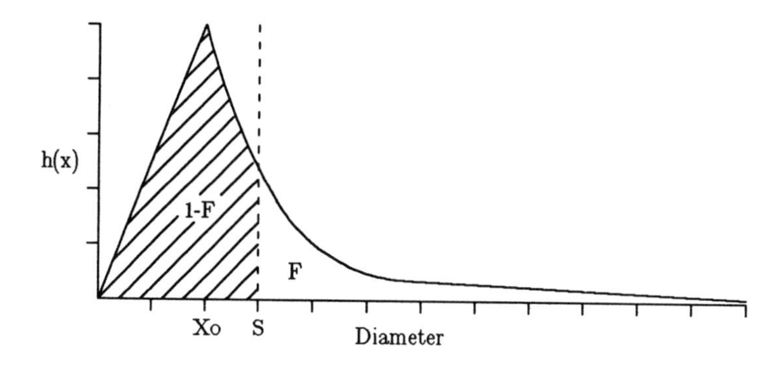

Figure 4-2: Truncated Defect Size Distribution

For minimum interesting defect size S, the fraction F of defects that must be generated is $x_0^2/(2S^2)$. If x_0 is 0.5 microns and S is 3 microns, then only 1 in 72 defects must actually be generated. We use this truncated size distribution to describe the diameter of extra and missing material defects. For this size distribution there is some chance of very large defects occurring. In practice these defects do not occur. We discard all defects larger than some maximum value x_M, typically several line widths in diameter.

4.2. Defect Spatial Distribution

Most of the early yield statistics research described in Chapter 2 concentrated on attempting to model how defects are spatially distributed across lots, wafers, and chips. Early models assumed that defects were randomly distributed with a uniform defect density. It was eventually recognized that defects tend to cluster within wafers and between wafers. This is due to the batch-oriented nature of the process, where conditions vary from lot-to-lot, and from wafer-to-wafer within a lot, due to time and space-varying particle concentrations in the air and chemicals. Conditions also vary across a wafer since the edge of the wafer is closer to the mechanical handling apparatus. It has been postulated that defect clusters are generated when vibration or other environmental changes cause a cloud of particles to break loose from the manufacturing equipment [Stapper 83a].

4.2.1. Lot and Wafer Distribution

Stapper assumed that defects within a wafer could be modeled by Poisson statistics and that the defect density between wafers followed a gamma distribution [Stapper 73]. Gamma distribution PDFs for several parameter values are shown in Figure 4-3.

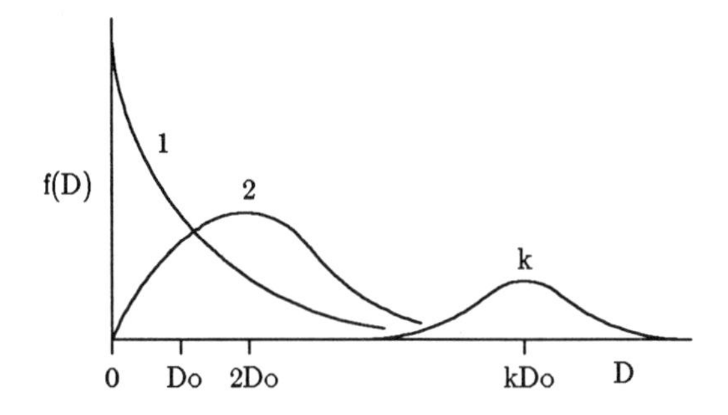

Figure 4-3: Gamma Defect Probability Density Function

The probability of finding x defects on a wafer is given by

$$\Pr(x) = \frac{\Gamma(\alpha + x)}{x!\,\Gamma(\alpha)} \frac{(\lambda/\alpha)^x}{(1+\lambda/\alpha)^{x+\alpha}} \tag{4-13}$$

where λ is the expected number of defects and α is the clustering coefficient. This is the negative binomial, Polya, or binomial waiting-time distribution [Johnson 69]. When α is an integer, the term Pascal distribution is used. In the case when $\alpha = 1$, the term geometrical distribution is used [Lukacs 72]. The geometrical distribution is the discrete analog of the exponential distribution. The negative binomial distribution is the limiting case of the Polya-Eggenberger distribution, and can be obtained from a modified form of Neyman's contagious distribution. Some examples of the negative binomial distribution are shown in Figure 4-4.

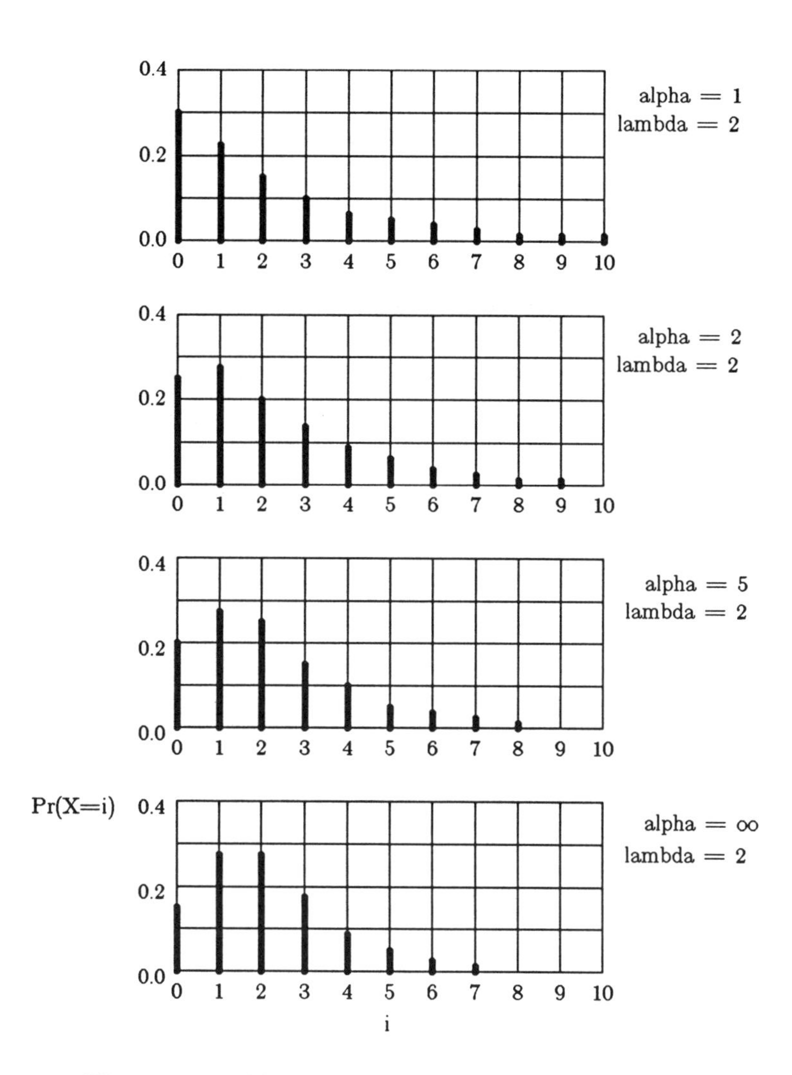

Figure 4-4: Negative Binomial Distribution Examples

The mean and variance of the negative binomial are

$$\bar{x} = \lambda, \tag{4-14}$$

$$\text{var}(x) = \lambda(1 + \lambda/\alpha). \tag{4-15}$$

The negative binomial distribution is often appropriate in cases where the observed variance is greater than the mean, as is often the case for local defects.

The average yield is

$$Y = \frac{1}{(1+\lambda/\alpha)^{\alpha}} \qquad (4\text{-}16)$$

Equation (4-16) provides a good fit to the data. In the case when $\alpha \rightarrow \infty$, the yield obeys Poisson statistics. In the case when $\alpha = 1$, the defect probability density function is exponential, with the yield given by Equations (2-13) and (2-15). Moore's half slice can be accurately modeled with negative binomial statistics [Stapper 75]. This model was later extended to multiple defect types [Stapper 82c] with the yield formula

$$Y = Y_0 \prod_i (1+\lambda_i/\alpha_i)^{-\alpha_i} \qquad (4\text{-}17)$$

where Y_0 is the yield after gross failures. This model has been used successfully at IBM [Stapper 82b, Stapper 85]. Stapper noted that for small numbers of lots, a variation in α_i was observed between lots, but that there was no interlot variation when averaged over longer periods of time. The value of α has been observed to decline over time, indicating greater clustering [Stapper 83c, Stapper 85]. Values of α were typically 1, but are now in the range 0.3. A beta mixing function was found to occasionally provide a better fit to the data, but was much more cumbersome to use, and so discarded.

Equation (4-16) has been used for design rule optimization, test coverage analysis, and defect density calculations, and yield modeling [Rung 81, Seth 84, Cleverley 83, Turley 74, Hemmert 81].

Paz and Lawson assumed that some fraction of the chip area was completely dead due to defect clusters [Paz 77]. The yield due to these defects was modeled by a beta distribution. The yield due to random defects followed a negative binomial distribution. This model was successfully applied to diffusion defects in bipolar transistor chains.

Almost all yield modeling work uses Murphy's Equation (2-7). This formula assumes that the defect density within a die is uniform. This is a good approximation when the size of a defect cluster is large compared to a die. However this condition is not necessarily the case for large chips or wafer-scale integration, leading to non-uniform defect densities. Stapper showed that negative binomial statistics still applies in this case [Stapper 84b, Stapper 86].

Armstrong and Saji found that Neyman Type A statistics provided a better fit to their data than Equation (4-17). The yield in this case is

$$Y = Y_0 \exp\left(-v + v\exp\left(-\sum_i A_{ci} D_i / v\right)\right) \tag{4-18}$$

where Y_0 is the yield after gross failures, A_{ci} and D_i are the critical area and defect density for defect type i, and v is the clustering parameter.

In this book, we use a two-level negative binomial distribution to describe the number of defects on a wafer. The first-level distribution models the variance between lots, and the second level models the variance between wafers within a lot. The distributions are described in Section 4.3. The composite negative binomial distribution has been rarely used [Johnson 69]. The only reported use in the literature is a two-level distribution to model the variance in the number of defects between wafers, and between chips within a wafer [Ketchen 85].

4.2.2. Defect Radial Distribution

A radial distribution in defect density across the wafer was observed [Yanagawa 69, Yanagawa 72, Gupta 72, Ham 78, Ferris-Prabhu 87]. Defects were usually found to be more common towards the edge of the wafer than in the center, due to the fact that these regions are more exposed to particulates in the air and are closer to the wafer handling apparatus. Gupta found the radial distribution to have the form

$$h(r) = c_1 + c_2 e^{c_1 r} \tag{4-19}$$

where r is the wafer radius, as shown in Figure 4-5 [Gupta 72]. The radial distribution was partly attributed to alignment error [Kim 78]. Others added an angular distribution in addition to the radial distribution, to

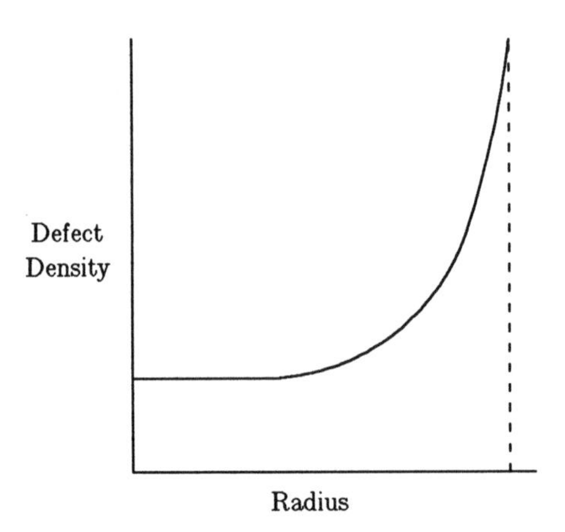

Figure 4-5: Radial Density Distribution

account for the fact that wafers have a flat, sit in boats, etc [Gupta 74].

Stapper analyzed the radial distribution by dividing the wafer into inner and outer zones [Stapper 76]. The defect density was normally higher in the outer zone due to defect clustering near the edge of the wafer. Paz and Lawson found the opposite to be true for their process [Paz 77]. Other researchers divided the wafer into as many as five concentric rings [Stapper 85].

We use a modified form of Stapper's two-zone radial distribution. The wafer is divided into a circular inner zone and concentric outer zone with different mean defect densities. Defects are assumed to be randomly distributed within a zone. The two-zone distribution is shown in Figure 4-6. We previously suggested the use of the piecewise linear distribution with a normal variance [Walker 85]. However such a model is not supported by the data in Chapter 8.

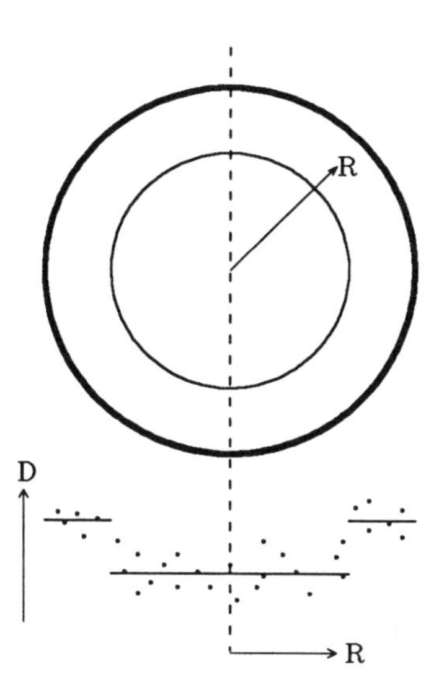

Figure 4-6: Two Zone Radial Distribution

4.3. Composite Defect Distribution

Based upon the above, we have decided on the combined defect spatial and size distribution as shown in Figure 4-7. For each chip sample, the number, size, and location of each type of defect is determined. For each defect type i, the number of defects $n_{l,i}$ in a lot is determined using the between-lot negative binomial distribution $NB_l(\lambda_{l,i}, \alpha_{l,i})$ where $\lambda_{l,i}$ is the expected number of defects per lot and $\alpha_{l,i}$ is the between-lot clustering coefficient for defect type i. These parameters are obtained using Equations (4-14) and (4-15). The expected number of defects per wafer within the lot $\lambda_{w,i}$ is calculated by the relationship

$$\lambda_{w,i} = \frac{n_{l,i}}{W}, \tag{4-20}$$

where W is the number of wafers per lot. The number of defects on the wafer $n_{w,i}$ is calculated using the between-wafer negative binomial distribution $NB_w(\lambda_{w,i}, \alpha_{w,i})$ where $\alpha_{w,i}$ is the between-wafer clustering coefficient. The values of the $\alpha_{w,i}$ are determined by applying Equations (4-14) and (4-15) to each lot, and averaging the resulting $\alpha_{w,i}$ values.

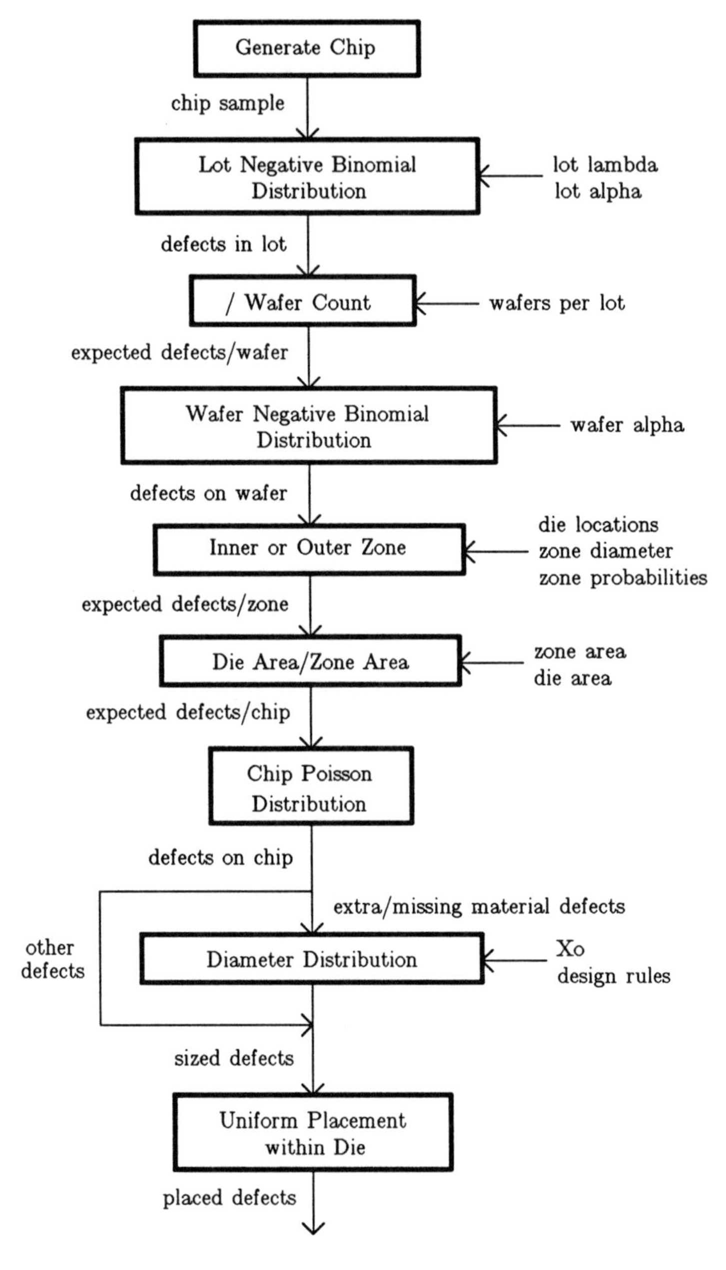

Figure 4-7: Combined Defect Spatial and Size Distribution

The chip lies in each zone of the wafer with probability

$$p_{iz} = \frac{c_{in,i}}{c} \qquad \text{inner zone}, \tag{4-21}$$

$$p_{oz} = \frac{c_{out,i}}{c} \qquad \text{outer zone}, \tag{4-22}$$

where $c_{in,i}$ and $c_{out,i}$ are the number of die in the inner and outer zones, and c is the number of die on the wafer. The values of $c_{in,i}$ and $c_{out,i}$ are determined from the zone diameter and the wafer map. The number of defects placed in the zones is

$$n_{in,i} = n_{w,i}\, p_{in,i} \qquad \text{inner zone}, \tag{4-23}$$

$$n_{out,i} = n_{w,i}\, p_{out,i} \qquad \text{outer zone}, \tag{4-24}$$

where $p_{in,i}$ and $p_{out,i}$ are the probabilities of defects occurring within the inner and outer zone respectively. The expected number of defects $\lambda_{c,i}$ on a chip is

$$\lambda_{c,i} = \frac{n_{in,i}\, a_c}{a_{in,i}} \qquad \text{inner zone}, \tag{4-25}$$

$$\lambda_{c,i} = \frac{n_{out,i}\, a_c}{a_{out,i}} \qquad \text{outer zone}, \tag{4-26}$$

where a_c is the chip area, and $a_{in,i}$ and $a_{out,i}$ are the inner and outer zone areas. The number of defects on the chip $n_{c,i}$ is then calculated using a Poisson distribution with mean $\lambda_{c,i}$. Diameters are then selected for the extra and missing material defects using the size distribution. The defects are placed uniformly within the chip. The implementation of the distributions is described in Chapter 6.

Chapter 5
Fault Analysis

This chapter describes the fault analysis phase of yield simulation. Once defects have been sized and placed on the chip layout, they must be examined to determine what circuit faults, if any, have occurred. Techniques for performing this analysis include local circuit extraction, layer combination analysis, and type-driven analysis. All of the fault analysis methods described here assume that the input layout geometry has been preprocessed to simplify analysis. We begin this chapter by describing this preprocessing phase. We then briefly describing the local circuit extraction and layer combination analysis methods and showing why they are not suitable for our purposes because of poor performance. We then describe the type-driven analysis method which is best suited for our purpose, and used in VLASIC.

5.1. Preprocessing

The fault analysis procedures assume that the layout has been preprocessed to identify and label transistors, and to label all net geometry with net numbers. This labeling is done with a circuit extractor, such as ACE [Gupta 81, Gupta 83a] or Magic [Ousterhout 85]. The layout is also processed to generate useful layer combinations, such as metal-polysilicon overlaps. The extracted layout is fully-instantiated, and divided into bins as shown in Figure 5-1. Each bin contains only a few polygons. The bins are used for locating layout polygons that lie in the defect neighborhood.

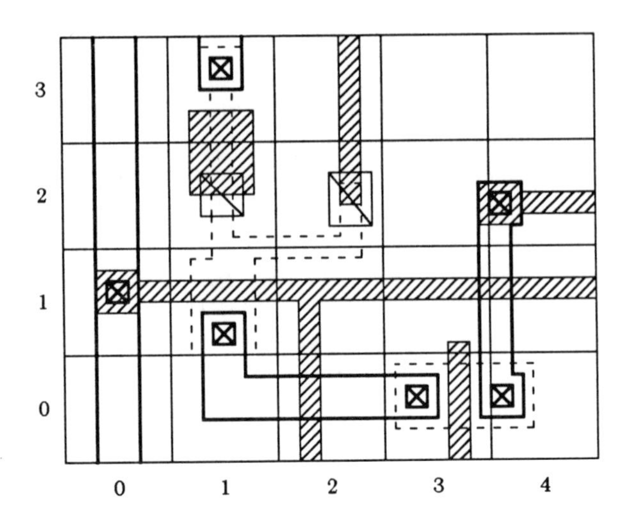

Figure 5-1: Layout Bins

5.2. Local Circuit Extraction

Fault analysis can be done by extracting a circuit from the layout in the neighborhood of the location of the defect both with, and without, the defect present, and then comparing these circuits. These two cases are shown in Figure 5-2.

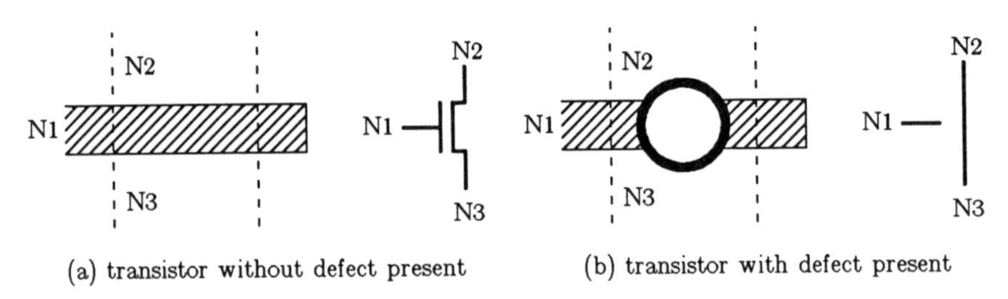

Figure 5-2: Neighborhood Circuit Extraction

If the two circuits are different, then a fault may have occurred. A more complex comparison is required to determine whether an actual DC change to the circuit topology has occurred, and the nature of the circuit fault or faults, such as a short or an open. Note that layout area beyond the actual location of the defect must be considered in order to provide a context for the detailed analysis.

This approach was implemented using the ACE circuit extractor to perform the circuit extraction in the neighborhood of the defect. Since ACE was designed for extracting layouts associated with large chips, it was modified to act as a subprocess to which layout geometry could be repeatedly passed, and circuits returned. Specifically, after a defect is placed on the layout, the geometry surrounding it is passed to ACE, and the circuits both with and without the defect present are returned.

If an extra material defect does not intersect any existing nets, then a new floating net is extracted. This means that the circuits with and without the defect are different, even though no circuit fault has occurred. In order to alleviate this problem, defects are specially labeled and any floating net which is extracted will have this label, and is discarded before the two circuits are compared.

One problem with using a standard circuit extractor to do the local circuit extraction is that it assumes that the input is design-rule correct. This is not necessarily the case if a defect is present. Second, the defects may not always be representable as geometry on the layout. For example, if an extractor does not handle buried contacts, there is no way to represent a gate oxide pinhole. Third, there is no way to represent a missing material defect to the extractor. The extractor must be modified to handle these situations. These modifications reduce extractor performance.

Since ACE only processes rectangles, non-rectangular polygons, such as the defect octagon, are approximated as a series of 0.5 micron tall rectangles. Hence each defect is represented by a large number of rectangles (21 for a 10 micron defect), so that circuit extraction with a defect present is much slower than circuit extraction without a defect present. The combined time for ACE to extract both the layout neighborhood with the defect present and the neighborhood without the defect present is typically 0.5 CPU seconds on a

VAX-11/780[1].

Once the two circuits have been extracted, they must then be compared with a graph isomorphism algorithm. Since the neighborhood around the defect is small, the circuits contain only a few transistors and nets, and the comparison takes only 10-50 msec [Ebeling 83]. To determine what fault has occurred, the two graphs are used to hash into a fault table. A similar technique has been used for diagnosing global defects [Odryna 85a, Odryna 85b].

In a typical yield simulation, hundreds of thousands of defects must be placed and analyzed. Even with a fast graph comparison and table lookup algorithm, analysis of each fault takes 0.3-0.5 seconds, which is unacceptably slow, requiring tens of CPU hours per simulation. Hence such an approach is unsuitable. However, a local circuit extraction approach to fault analysis could potentially be made much faster by constructing a circuit extractor optimized for this application [Chew 87]. This possibility is discussed further in Chapter 9.

5.3. Layer Combination Analysis

The fault analysis problem can also be approached by considering the chip layout as a set of polygonal tiles. Each such tile represents some layer combination, such as metal1 and poly, metal1 and metal2 and second-level via, or active and poly. Each of these layer combinations has an electrical meaning. Active and poly form a transistor. Metal1 and poly form parallel conductors. Metal1 and metal2 and second-level via form a conductor between metal1 and metal2. These layer combinations and meanings can be automatically generated from a process description [Maly 84a, Shen 85]. Each tile therefore represents an electronic component. By using adjacency information and electrical connectivity rules (e.g. conductors on the same layer connect if adjacent), these electronic components can be combined to

[1] VAX is a trademark of the Digital Equipment Corporation.

form the circuit.

A defect can modify the layer combinations of those tiles intersecting it, and hence their electrical meaning, resulting in a circuit fault. When a defect is placed on the chip, the tiles in the defect neighborhood are examined without the defect present. Their electrical meaning and adjacency are used to determine the unmodified neighborhood circuit. The defect is then added to the layout. In the case of a defect modeled as a point, such as an oxide pinhole, a very small octagon is placed on the layout. The defect is subtracted from the neighborhood tiles. Those tiles and parts of tiles that remain are unmodified. Since the tiles and defect are polygons, the subtraction is done with polygon operations. The defect is then intersected with the neighborhood tiles. The resulting tiles lie under the defect. Their layer combinations are modified according to the defect type, based on a table lookup. For example, a missing metal1 defect removes metal1 from all tiles under the defect. The adjacency of the newly created tiles is determined through the use of a special polygon adjacency test. The tiles are then converted to electronic components and combined with the adjacency information to construct the modified neighborhood circuit. The two circuits are then compared as in the previous section. The layer combination analysis essentially replaces the circuit extraction step in the previous section with a table-driven extractor.

As an example, an extra active defect is placed on the layout so that it intersects two active lines. The layout is partitioned into tiles with and without the defect present, as shown in Figure 5-3. Note that a large number of tiles may result. Without the defect, tiles 1-5 form a metal1 line N1, and tiles 6, 2, 7 and 8, 4, 9 form active lines N2 and N3. With the defect added, tiles 1-5, 10, 11 form a metal1 line N1, and tiles 2-4, 6-13 form an active line N2-N3, created by shorting the two original active lines N1 and N2 together.

The advantage of the layer combination fault analysis technique is that it can be readily table-driven and handles arbitrary geometry. The

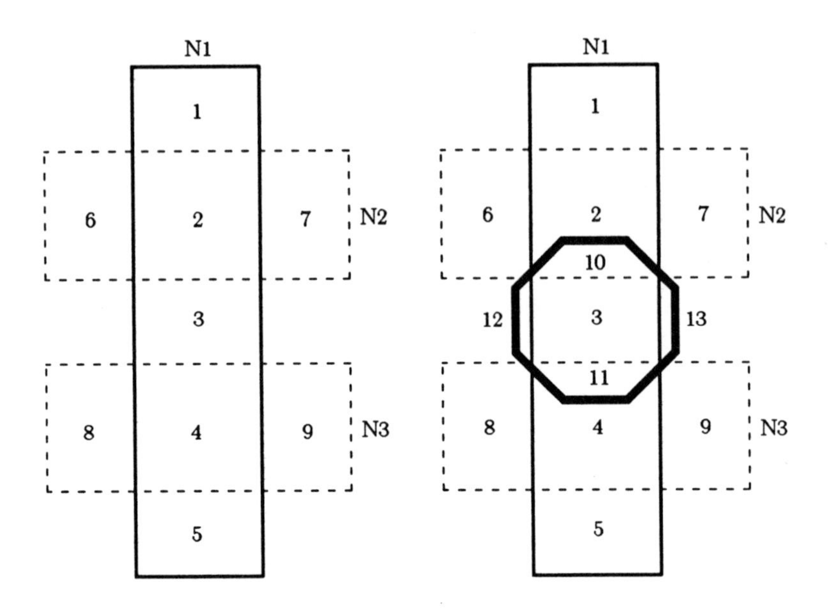

Figure 5-3: Layer Combination Analysis

disadvantage is the need to chop up the tiles when the defect intersects them, and to determine how the new tiles are interconnected. This requires a large number of polygon operations. The operation of splitting the tiles and then determining adjacency places heavy demands on the accuracy of the polygon operations. In addition, the newly-created tiles have many edges. These effects combine to produce a slow implementation, so the layer combination approach to fault analysis does not appear promising. However if the layout is restricted to Manhattan geometry, acceptable performance may be obtained [Ferguson 87].

5.4. Type-Driven Analysis

Fault analysis can also be organized around the types of fault that occur in the circuit. Note that each defect type can only cause some fault types. For example, an extra metal defect can only cause a short circuit. As described in Chapter 3, there are only a limited number of circuit fault typesthat can cause a change to the DC circuit topology. The complete list of such fault types is:

- *Short* - a short circuit between conductors on the same layer, or through an incomplete via, due to an extra material defect.

- *Open* - an open circuit due to an open via or broken conductor, due to an extra or missing material defect.

- *Shorted Device* - a device with source shorted to drain due to a missing material defect.

- *Open Device* - a device with an open circuit in the channel due to a missing material defect.

- *New Via* - a short between conductors on two different layers caused by an oxide pinhole or junction leakage defect.

- *New Gate Device* - the creation of a new device, the creation of a multi-terminal device, or the addition of a terminal to an existing device due to an extra gate (poly) material defect.

- *New Active Device* - the creation of a new device, the creation of a multi-terminal device, or the addition of a terminal to an existing device due to an extra active material defect.

Type-driven analysis uses a set of special-purpose fault analysis procedures, corresponding to the fault types listed above. These procedures embed the knowledge of what geometry configuration is necessary to cause that type of circuit fault. When a defect is placed on the layout, a process table lookup is performed to determine what types of faults the defect may cause. The analysis procedures for each of these fault types is called to examine the neighborhood of the defect to determine whether any of that type of fault have occurred. As discussed in Section 5.1, the chip layout is divided up into rectangular bins, with all the geometry intersecting a bin stored in that bin. This allows layout geometry neighboring the defect to be readily determined. Neighboring geometry is defined as all the geometry in all the bins touched by the defect. An example is shown in Figure 5-4 with the defect touching the bins (0, 0), (0, 1), (1, 0), and (1, 1). The process tables also record what layers interact with the particular defect type. For example, an extra metal

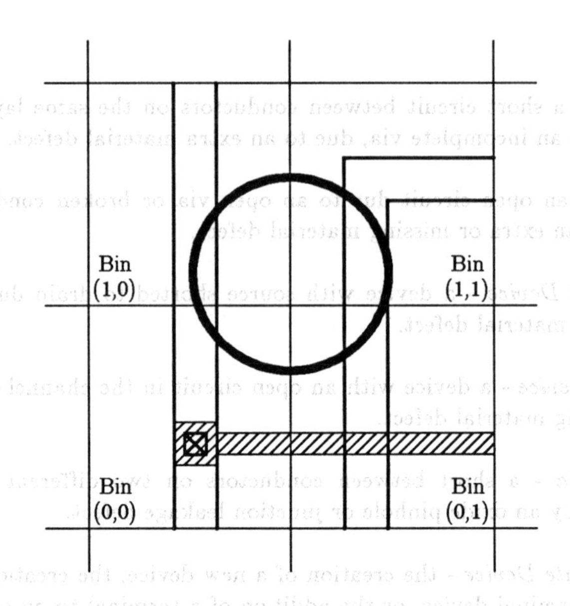

Figure 5-4: Geometry in Neighboring Bins

defect only interacts with the metal and metal via layers. Since the layout geometry consists of polygons, general polygon operations are used in the fault analysis procedures.

The fault analysis procedures are greatly simplified if polygons in the neighboring bins are *merged* before use. Merging is done by unioning together all polygons that touch one another, as shown in Figure 5-5.

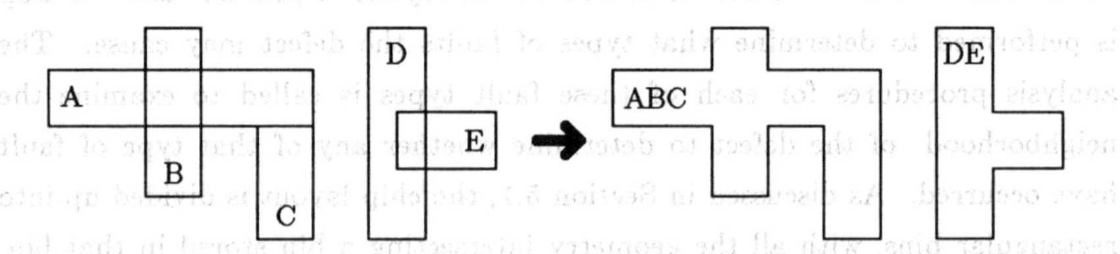

Figure 5-5: Polygon Merging

Polygons are only merged in the defect neighborhood, rather than across the entire chip, to avoid the creation of very large polygons, that can be expensive to process. This point is discussed further in Chapter 6.

To maintain generality, fault analysis procedures make no assumptions about the shape of the defect polygon. The models in Chapter 3 describe defects as octagons or points, but this fact is only taken advantage of in the new via analysis procedure. Arbitrarily shaped defects require more complex fault analysis procedures than would be the case if the defect was assumed to be an octagon. In general, convex defects are easier to analyze than non-convex defects.

The analysis procedures may discover more than one fault. Several of these faults may have to be combined or discarded. For example, the New Gate Device procedure may determine that an extra poly defect has caused the creation of a new device while the Short procedure may have determined that the extra poly is shorted to some other net. This short circuit must be propagated to the new device by connecting the gate of the new device to the net. The fault combination may result in a fault that is not a DC topology change, and hence must be discarded. For example, a new device may be detected by one procedure, while another procedure determines that the source will be shorted to the drain. This device then forms a capacitor, and so is discarded. Most new via faults are electrically equivalent to short circuits, so they are converted to shorts. As discussed in Chapter 3, a new via due to an oxide pinhole is not converted to a short.

Type-driven analysis has the advantage that the average fault analysis can be done quickly. The first two fault analysis methods suffered from using overly general techniques. Most circuit faults result from relatively simple configurations of defect and layout. For example, a short occurs if an extra metal defect touches two or more metal polygons with different net numbers. This can be easily determined with some polygon intersection tests between the defect and neighboring metal polygons. Complex fault analysis need only be done in those rare cases where a complex fault may occur.

Another advantage of using type-driven analysis is that the procedures know what they are looking for. There is no need to perform a general graph isomorphism or pattern recognition step. For each fault type, only a few graph transformations can occur, and these are built into the analysis procedure. Because the analysis procedures are only looking for a *change* in the circuit, there is no need to examine the layout with and without the defect present.

The following sections describe the fault analysis procedures for detecting shorts, opens, new vias, open devices, shorted devices, new gate devices, and new active devices. The fault combination and filtering procedure is then described.

5.4.1. Shorts

A short circuit occurs when an extra material defect polygon touches two or more neighboring polygons on interacting layers on different nets (having different net numbers). An interacting layer is one that can electrically connect to the defect material, as defined by the defect models. For example, an extra metal defect can electrically connect to metal lines, while an extra poly defect can electrically connect to other poly, and to active through a buried contact, as was shown in Figures 3-3 and 3-4. If a short circuit occurs, the short procedure returns a list of the shorted net numbers.

The touch operation is implemented as a polygon intersection where only a boolean result is computed, rather than the actual intersection polygon. The net numbers of all polygons touching the defect are recorded. This list of net numbers is examined to remove duplicates. If two or more unique net numbers remain, then a short has occurred between these nets.

There are two complications to the short analysis procedure. First, shorts caused by extra active or gate (poly) material must be recorded even if they only touch one net. This information will later be used to assign net numbers to the gate and source/drain terminals of new gate or active devices. These one-net shorts are then discarded.

The second complication is that some layers mask others. In a MOS process with self-aligned transistors, the gate (poly) layer masks the active layer, except where the gate intersects the buried contact layer. Where gate overlaps active, a transistor channel occurs instead of a diffusion line. Where the gate intersects a buried contact, a poly-active contact is formed. Therefore when considering shorts due to extra active material, the defect polygon must first have neighboring gate polygons that do not lie over buried contact subtracted from it. This situation is shown in Figure 5-6. If masking is not done, a N1-N2-N3-N4 short will be reported, rather than shorts N1-N2 and N3-N4.

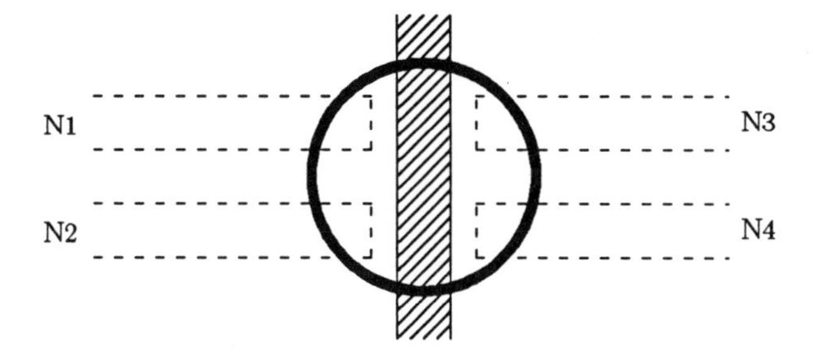

Figure 5-6: Masked Active Material Defect

This masking operation is also needed by the new active device analysis procedure, so the work is shared between them.

5.4.2. Opens

An open circuit occurs when a missing material defect spans a line or via on an interacting layer, splitting a net into several unconnected branches. An interacting layer is one that can be broken by the missing material. For example, a missing metal defect can break a metal line, or a missing first-level via defect can open a first-level via. The defect models separately record whether a defect type can span lines or cause open vias. An example open line and open via were shown in Figures 3-12 and 3-13. If an open circuit occurs, the open procedure returns the list of branches for each broken net. Each branch contains a record of all transistor terminals and bin boundaries attached to the branch in the defect neighborhood, as shown in Figure 5-7.

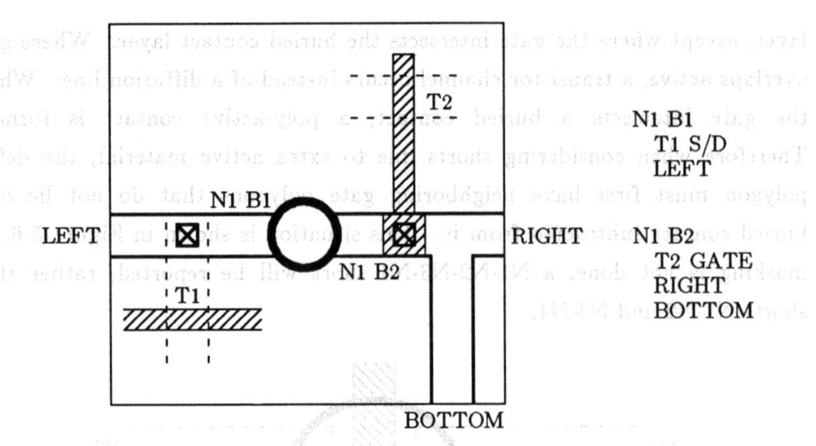

Figure 5-7: Branch Records

This branch information can be used by post-processors to determine exactly where in a net graph the open circuit has occurred. There are several reasons that we do not determine the exact location of an open circuit during fault analysis. First, most applications do not need this information. It is often sufficient to locate the fault within a set of bins. Second, this information is potentially expensive to provide. The obvious technique for exactly locating an open circuit is to list all transistor gates and terminals in each branch. Since most nets are small, and connect only a few transistors, this analysis does not take much time for most broken nets. However analysis of many circuits shows that a significant number of nets, such as power and clock lines, have very high fanout [Frank 85]. Locating all of the transistors on each branch of these nets would be very time consuming, using the techniques described below. It amounts to circuit-extracting each branch of the broken net. In addition, these nets consume a large fraction of total net length, and hence are more likely to be broken by a defect. The combination of high fanout and high probability of breakage means that the open circuit location time would be dominated by these large nets.

The expense of locating an open circuit can be greatly reduced by preprocessing the circuit graph to mark all those places in the graph where a net crosses a bin boundary. Given the open location within a bin, and where the branches touch the bin boundaries, it is possible to locate the open in the entire graph. Even with this improved method, determining the exact open location would probably be too expensive, so we leave it to a post-processor.

The first step in analyzing an open circuit is to mask the defect, to remove those parts that cannot cause opens. The only case of masking in a MOS process is removing the intersection of active and buried contact from an extra gate (poly) defect as shown in Figure 5-8.

Figure 5-8: Masked Extra Gate Defect

The intersection of these three layers forms a contact, and so an open circuit cannot form. If after masking, there is no defect left, then an open circuit cannot occur.

Each interacting polygon that can be spanned is subtracted from the masked defect. If the number of contours in the result is greater than in the masked defect, then the masked defect is subtracted from the polygon. If this subtraction results in several contours, then an open circuit has occurred. This sequence of operations is shown in Figure 5-9.

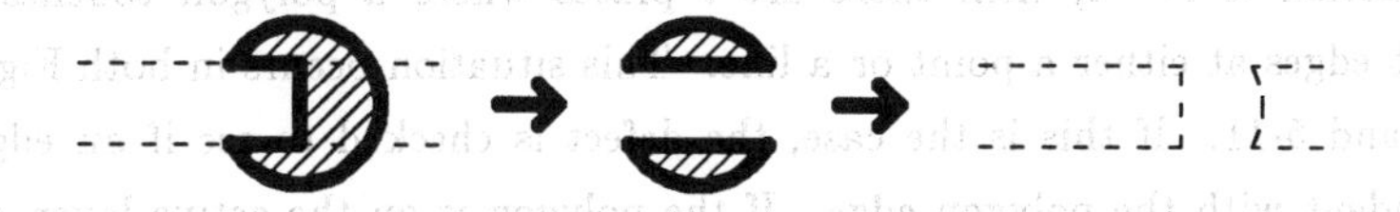

Figure 5-9: Polygon Operations for Open Detection

The two subtractions are used to deal with some problems that occur if a defect edge and polygon edge are coincident. This is an artifact of using octagonal defects, rather than circles. The defect octagon could be at the end of a line, as shown in Figure 5-10. If an open was detected by subtracting

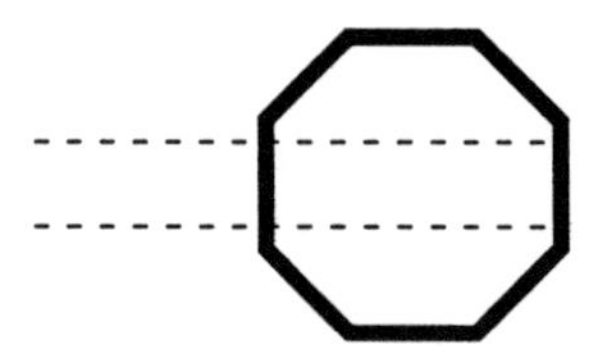

Figure 5-10: Defect Coincident with Polygon Edge

the defect from the polygon, then in this case, only one contour would result, and the open would be missed. If the polygon is the end of a line, this is no problem since no open has occurred. However the polygon could be an active area connected to a transistor source/drain terminal as shown in Figure 5-11.

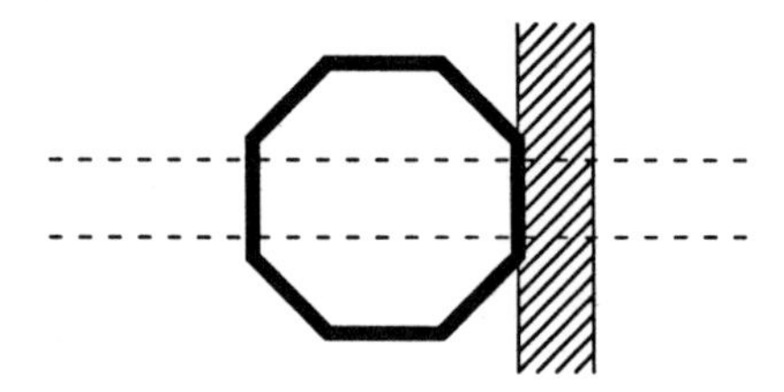

Figure 5-11: Defect Coincident with Transistor Edge

This is defined as an open circuit. In order to detect the open, we subtract the active polygon from the defect first. If the defect is split into two or more contours, we then subtract the defect from the polygon. If the contour count of the first subtraction is N and the contour count of the second subtraction is $N-i$, then there are i places where a polygon touches the defect edges at either a point or a line. This situation occurs in both Figures 5-10 and 5-11. If this is the case, the defect is checked to see if an edge is coincident with the polygon edge. If the polygon is on the active layer, then the the polygon edge is either the end of a line, or a transistor terminal. A coincidence check is done with all transistor channels in the neighborhood. If a transistor is coincident, then the open is valid, otherwise the open is discarded. This coincidence situation can only occur if one edge of the defect octagon is at least as long as the end of a line. For lines of width W, the defect diameter must then be at least $2.41W$.

If an open circuit has occurred, the contours of the broken polygon are decomposed into individual polygons as shown in Figure 5-12.

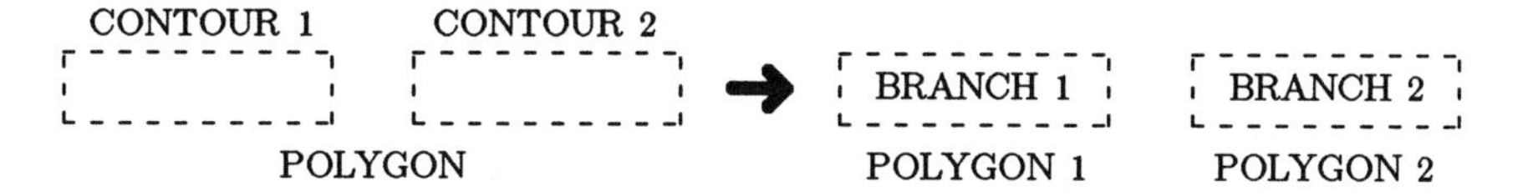

Figure 5-12: Decomposed Multi-Contour Polygon

These new polygons represent the starting point of each branch of the broken net.

In the case of a break coincident with a transistor terminal, there is no active polygon to represent the branch containing the transistor. Such a polygon is required by the branch traversal algorithms described below. This problem is solved by generating a dummy sliver polygon to represent this branch.

When subtracting the defect from a net polygon, the polygon may be completely deleted. This can happen in only three cases. In the first case, the polygon is not connected to the rest of the circuit, and so is ignored. In the second case, the polygon is on the active layer connected to transistor source/drain terminals. The fact that the missing active material defect surrounded the polygon means that it also removes part of the transistor channels, and so is detected by the open device analysis procedure. In the third case, the polygon is connected to other layers through a via, and so an open will be detected by the open via analysis described below. In all three cases, if the polygon is completely deleted, it is not regarded as a broken net.

If the defect type can cause open vias, then it is subtracted from interacting polygons. Note that the defect polygon is used, not the masked defect polygon. We assume that open vias cannot be masked by other layers. If the result of the subtraction is a null polygon, then the defect completely covers the via, and an open via results, breaking a net, as shown in Figure 5-13. The top and bottom layers connected to the open via represent

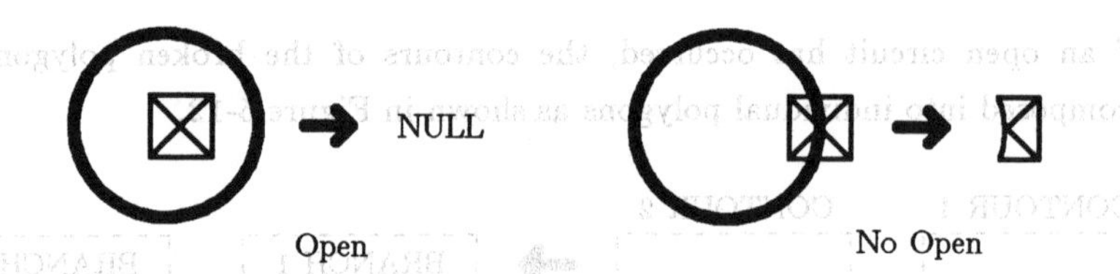

Figure 5-13: Open Via Detection

starting points for the net branches. If no open vias or spanned polygons have resulted from the above analysis, then no open circuits have occurred.

The branch starting points for spanned polygons are used to initiate a recursive traversal of all polygons electrically connected to the branch. For each polygon, the transistor gates and terminals touching it are recorded. Then the bin boundaries touched by the polygon are recorded. The traversal does not extend beyond the bins containing the defect. Then the traversal recursively examines all those polygons connected to the current polygon through a via. The process table determines what layers each via type connects to. Each polygon is marked as it is visited. Since polygons on each layer are merged, there is no need to check for the polygon touching another polygon on the same layer. The traversal algorithm dependents on the assumption that there is only one active and one gate (poly) layer. An example traversal is shown in Figure 5-14.

For the case of extra material defects, several branches can be reconnected by shorting through the defect, as shown in Figure 5-15. In a MOS process, this can only happen by buried contacts reconnecting broken active lines through an extra poly defect. If the possibility of this reconnection exists, the branch traversal algorithm checks for a via (a buried contact in this case) connecting a polygon to the defect. This information is used to perform the reconnection in the fault combination phase.

Nets with loops present a problem to the branch traversal algorithm. When a polygon is spanned by the defect, the broken pieces are not stored in the layout data structures (the geometry bins). To do so would be quite

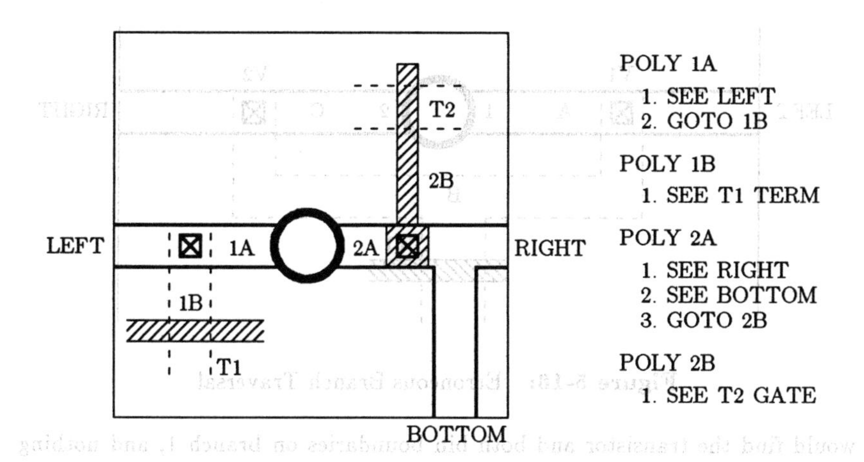

Figure 5-14: Branch Traversal

Figure 5-15: Reconnected Branches

expensive. Instead, a separate list of pieces is kept. Each of these pieces is used as a starting point for the branch traversal. The original unbroken polygon is left in the layout structure. In order to prevent the branch traversal from visiting it, it is marked as visited. However if a loop exists in the broken net, this marking will lead to an erroneous traversal. An example of this is shown in Figure 5-16. Branch 1 first visits polygon A, seeing the left bin boundary. The traversal then goes through via V1 to polygon B, and sees a source/drain terminal of transistor T1. The traversal then attempts to go through via V2. However the polygon on the other side of the via is the original unbroken metal1 polygon, which has been marked as visited. Therefore the traversal halts. Branch 2 visits polygon C, seeing the right bin boundary. When the traversal attempts to go through via V2, it finds that polygon B has already been visited, and so halts. The result of the traversals is that branch 1 sees the left boundary and a source/drain terminal of transistor T1, while branch 2 sees the right boundary. The correct traversal

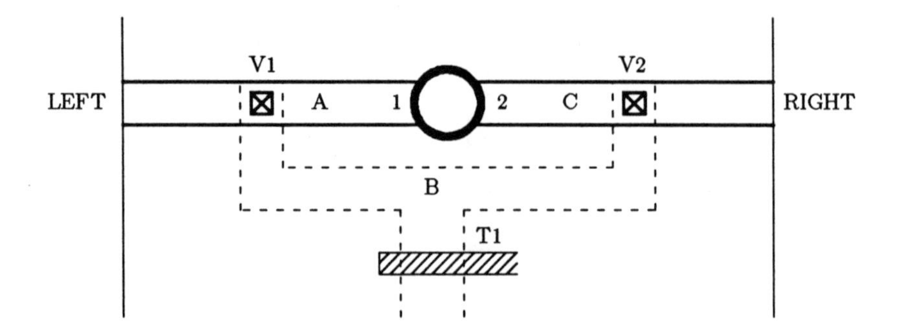

Figure 5-16: Erroneous Branch Traversal

would find the transistor and both bin boundaries on branch 1, and nothing on branch 2. As is discussed in the fault combination section, the fact that branch 2 has no elements means that the open circuit is discarded, which is the correct action.

One possible way to detect the above situation is for the traversal to check the visited polygon that blocks the traversal through a via. If that polygon is the original unbroken polygon, then the broken polygons would be checked for traversal. However the algorithm is designed so that broken polygons must be the first polygon in a traversal. Modifying it to check for loops would greatly complicate an already complicated algorithm.

Circuits loops occur frequently in some processes, in particular, the double-metal NMOS process considered here. The most common case is when polysilicon control lines have been "stitched" with metal. A metal line is run parallel to the poly line, contacting it periodically. This lowers the series resistance of the control line to approximately that of the metal, instead of the poly. Loops can also occur in power nets, especially in the input/output pads. These loops usually occur over an area much larger than the defect neighborhood. In this case, the local analysis will report that an open circuit has occurred, with the branches exiting the neighborhood. An application post-processor is left to determine whether the circuit contains a loop or not.

A problem in the branch traversal can occur even if there are no loops. A defect may break a net in two places, as shown in Figure 5-17.

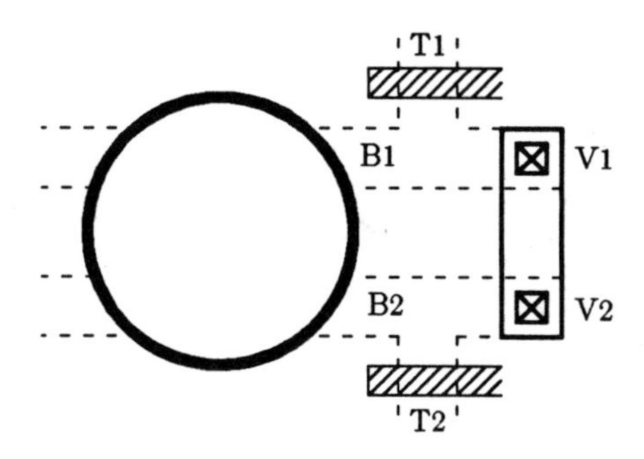

Figure 5-17: Double Net Break

Branch B1 will see a source/drain terminal of transistor T1, and then pass through via V1 to via V2. The traversal must terminate there because the polygon on the bottom of V2 is broken, and so already marked as visited. The traversal starts again with branch B2, which sees transistor T2, and is blocked at via V2, which is already visited. Two branches result, when in fact there is only one branch. The solution proposed for dealing with loops applies here also.

After the spanned polygon branches are traversed, the unvisited branches of open vias are traversed. As is the case for spanned polygons, the open vias must be marked as visited in the layout so that the traversal does not use them. The broken polygons must be examined before the open vias because they are not located in the layout. If the conductors on the top and bottom of an open via were traversed first, this might isolate a broken polygon on some branch, similar to the case with loops. An example of this is given in Figure 5-18. If the polygons A and B on the top and bottom of open via V1 are traversed first, branch B1 will see the top boundary, and branch B2 will see transistor T1, but then halt at via V2. Branch B3 starting from the broken polygon C will halt at via V2. This incorrectly divides the branch containing transistor T1 into two branches B2 and B3. If the traversal starts with polygon C, then transistor T1 will be seen, and traversal will halt at

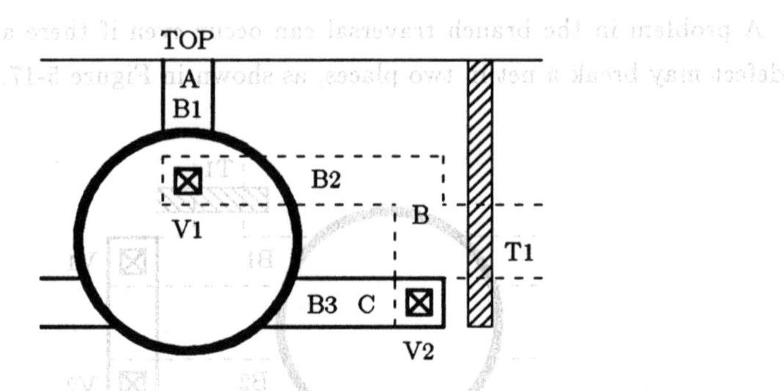

Figure 5-18: Problem Traversing Open Vias First

open via V1. This correctly combines branches B2 and B3 into one branch.

The location of an open circuit is identified by the transistors and bin boundaries on each branch. A problem arises if a missing gate (poly) material defect removes a transistor gate. It may be the case that no branch will see the transistor, as shown in Figure 5-19.

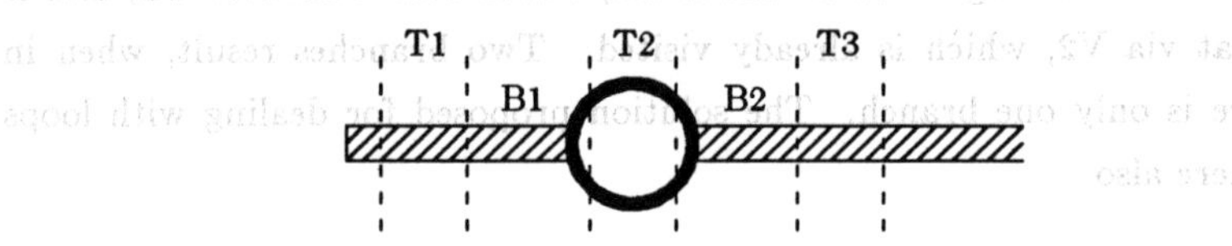

Figure 5-19: Open Location Uncertainty

Therefore it is not possible to directly locate the break in the circuit graph. The open circuit can be located from context. Branch 1 sees the gate of transistor T1. Branch 2 sees the gate of transistor T3. Therefore there is at least one break in the line connecting T1 and T3. By knowing the order that transistors T1, T2, and T3 appear on the gate polygon, it is possible to deduce that both the line between T1 and T2, and the line between T2 and T3 are broken.

5.4.3. New Vias

A new via occurs when an oxide pinhole defect intersects a region where lines on conducting layers above and below the oxide layer overlap, as specified by the defect models. These overlap areas are precalculated during the preprocessing phase. New vias also occur when a junction leakage defect intersects a diffusion region (active minus channel). Since the defects are always assumed to be points, the intersection test can be done by calculating the winding number of the point with respect to the neighboring overlap polygons. This is much cheaper than other intersection tests. If a new via occurs, the new via procedure returns the pair of nets shorted together.

New via faults are distinguished from shorts for two reasons. First, a new via caused by a gate oxide pinhole does not have a well-defined electrical meaning, and we would like to distinguish it from a short circuit. Second, the fact that oxide pinholes and junction leakage defects are always assumed to be points makes their analysis much easier.

5.4.4. Open Devices

An open device occurs when a missing active area defect spans a transistor channel from side to side, thus breaking the channel between source and drain, as was shown in Figure 3-14. If an open device occurs, the open device procedure reports the device number.

An open device is detected by intersecting gate (poly) polygons with the defect, isolating the defect under the gate. The result may have several disjoint contours, which are decomposed into separate isolated defects, called masked defects, which are regarded separately. Several masked defects are created if a single defect touches more than one transistor, as shown in Figure 5-20. If no masked defect results from the intersection, then there can be no open transistors. Channel polygons are subtracted from the masked defect. If the result has two or more contours, then the channel has been spanned by the defect, resulting in an open transistor, as shown in Figure 5-21. The channel is subtracted from the defect, rather than the defect from

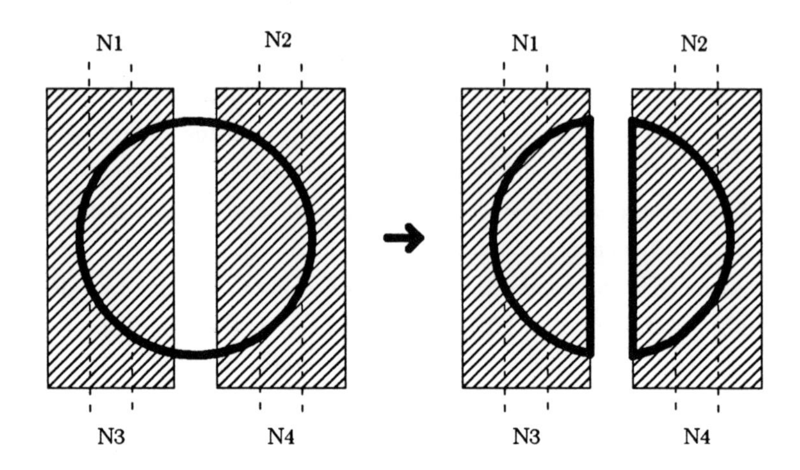

Figure 5-20: Multiple Open Transistors

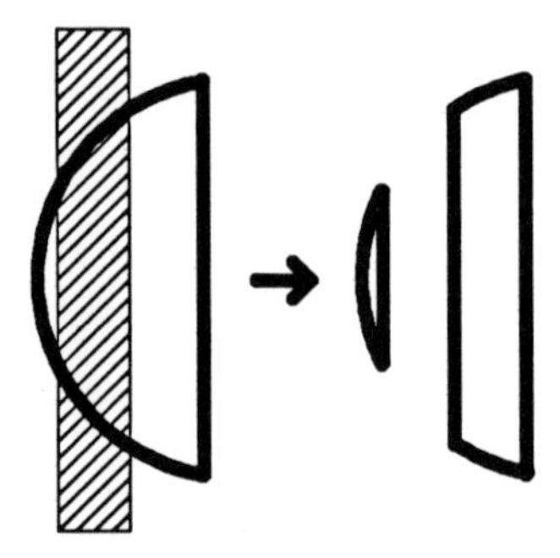

Figure 5-21: Open Device Detection

the channel, in order to avoid missing an open device when the defect and channel edges are coincident, as shown in Figure 5-22. Just subtracting the defect from the channel would result in only one contour, so the open device would be missed. This situation is analogous to the situation for open circuits at a transistor source/drain.

The open device procedure assumes that the gate (polysilicon) extends beyond the side of the transistor channel. This is normally required in all fabrication processes. If the gate edge were only coincident with the channel edge, then the analysis will miss some open devices, as shown in Figure 5-23.

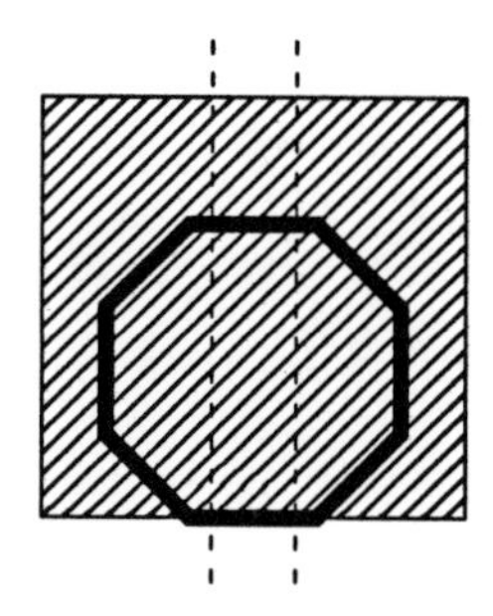

Figure 5-22: Defect Coincident with Channel Edge

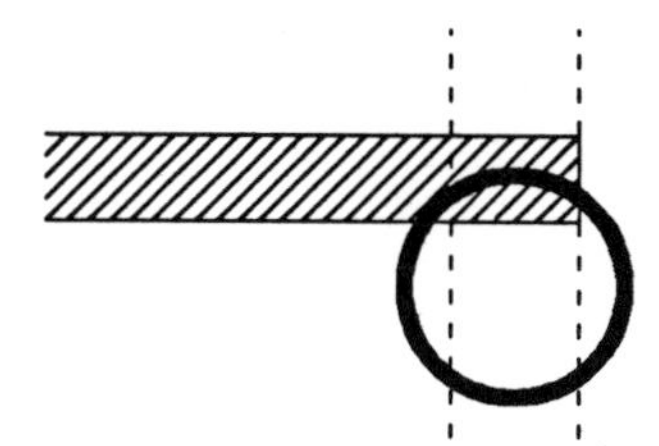

Figure 5-23: No Gate Extension Beyond Channel

In this case, when the defect is isolated over the channel, and the channel subtracted from it, only one contour results.

5.4.5. Shorted Devices

A shorted device occurs when a missing gate (poly) defect spans the channel all the way from source to drain, allowing diffusion material to short between source and drain, as was shown in Figure 3-15. If a shorted device occurs, the shorted device procedure reports the device number.

The active polygons provided by the circuit extractor have the channel regions subtracted from them, with the channel polygons provided on a special channel layer. In order to perform the shorted device analysis, the complete active mask must be recreated by taking the union of the channel polygons and active polygons in the defect neighborhood. The defect is intersected with the complete active polygons, creating a masked defect, as

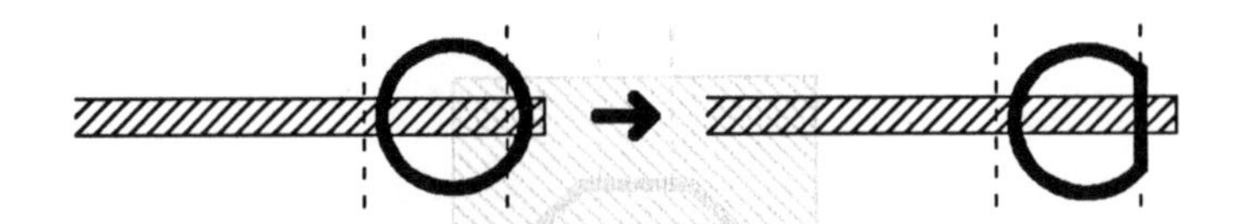

Figure 5-24: Masked Defect

shown in Figure 5-24. The channels are subtracted from the masked defect. If the result has two or more contours, then the transistor is shorted, as shown in Figure 5-25.

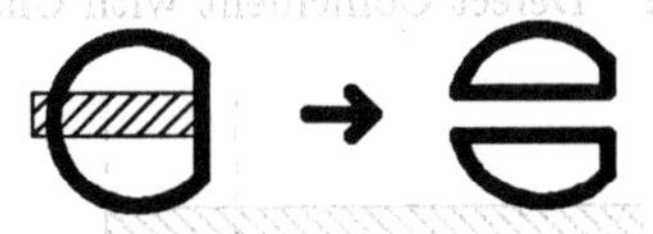

Figure 5-25: Channel Subtracted from Defect

The gate must be subtracted from the masked defect rather than vice versa for the following reason. If the defect covers one of the side edges of the channel, as in Figure 5-24, then the masked defect shares an edge with the channel. If the masked defect is subtracted from the channel, the channel will not be split in two, but only have its end removed.

The defect is isolated over the channel because in fact a single missing gate (poly) defect can cause more than one transistor to be shorted, as shown in Figure 5-26.

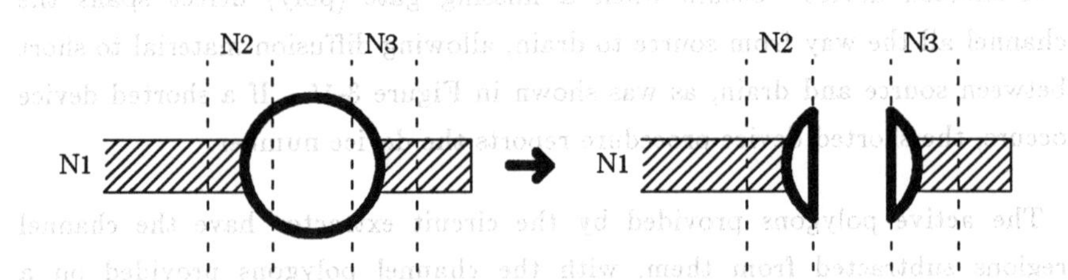

Figure 5-26: Multiple Shorted Transistors

When the defect is masked, the result may have several contours, corresponding to intersections with several channel/active regions. These contours are decomposed into several polygons, and considered as several individual masked defects that can cause several different shorted transistor faults.

The shorted device procedure assumes that the complete active polygons extend beyond the gate, and are not coincident with the channel edge. If such a coincidence occurs, then the channel may be shorted from the source to drain edge without being detected. This situation is shown in Figure 5-27.

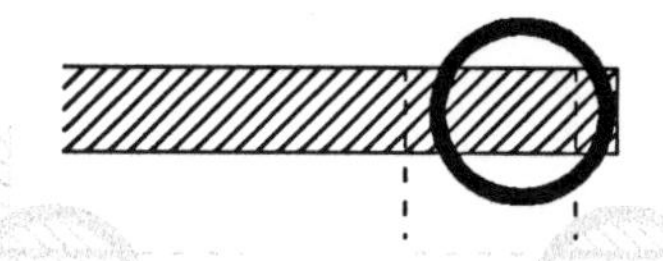

Figure 5-27: Active Edges Coincident with Channel Edges

Other more complex geometry operations can be used to avoid this problem. A shorted device can be detected by looking for a sequence of defect polygon edges that lie under the channel, whose exterior sides are not coincident with a channel side edge, and that touch both source and drain edges at some place other than their endpoints. Since all transistors must normally have active regions extend beyond the channel, this more complex analysis is not performed.

5.4.6. New Gate Device

A new gate device occurs when an extra gate (poly) defect spans an active line, breaking it into two or more pieces, and creating a channel connected to the source and drain terminals, as was shown in Figures 3-6 and 3-9. A new gate device can also occur by converting an existing two-terminal transistor into a multi-terminal transistor by adding new source/drain terminals, as was shown in Figure 3-10. If a new device occurs, the analysis procedure reports a new device with source and drain terminals connected to branches of a broken active net. If the new device is converted from existing transistors, the names of the transistors involved are also reported. The gate is connected in the fault combination phase.

The first step in the analysis procedure is to subtract the intersection of buried and active from the defect, creating a masked defect, as was shown in Figure 5-8. These regions are subtracted since they form poly-active contacts rather than transistors. As was the case for the shorted device procedure, the complete active polygons must be recreated by unioning the active and channel polygons. The masked defect is intersected with these complete active polygons, creating new channel regions. These channel regions may include the channels of existing transistors. Both situations are shown in Figure 5-28.

Figure 5-28: New Channel Regions

If no new channel polygons are created, then a fault cannot occur.

The masked defect is subtracted from the complete active polygons, breaking apart those polygons that form the transistor source/drain terminals, as shown in Figure 5-29.

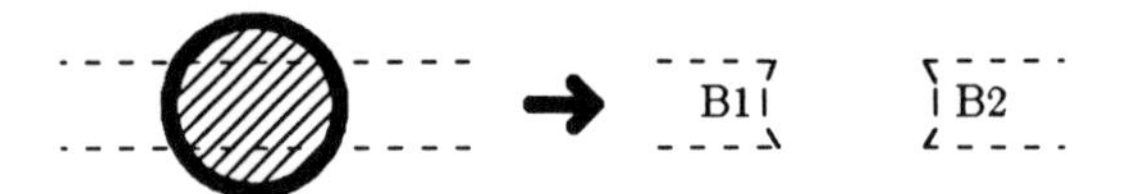

Figure 5-29: Defect Breaking Active Polygons

These broken polygons are given the same branch numbers as used by the open circuit fault analysis procedure. These numbers will be used in the fault combination phase to combine new device and open circuit faults.

Each new channel polygon is unioned with all existing channels that touch it, creating a combined channel polygon. The existing channels that touch are recorded. The net and branch numbers of all broken active polygons that touch the combined channels are recorded on the combined channel terminal list. The terminals of existing channels participating in the combined channel are added to the list, if not already present. These steps are shown in Figure 5-30.

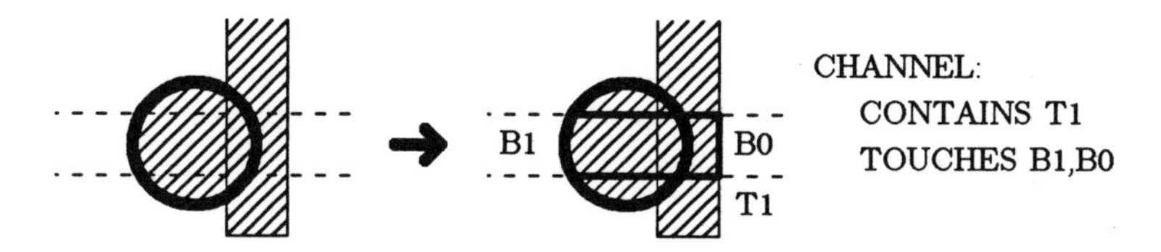

Figure 5-30: Channel Creation and Connection

Those source/drain terminals that were not created by breaking an active polygon represent the terminals of existing transistors. They are assigned branch number 0.

A new device is created in the following cases:

- The combined channel has two terminals, and does not contain any existing channels. This is a new transistor.

- The combined channel has more than two terminals, and does not contain any existing channels. This is a new multi-terminal transistor.

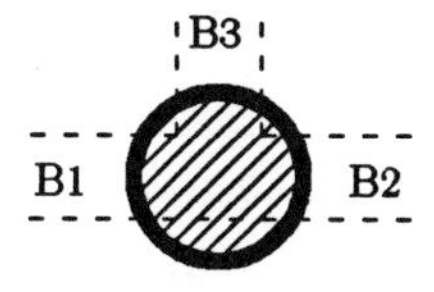

- The combined channel has more than two terminals, and also touches existing channels. This is a multi-terminal transistor converted from existing transistors.

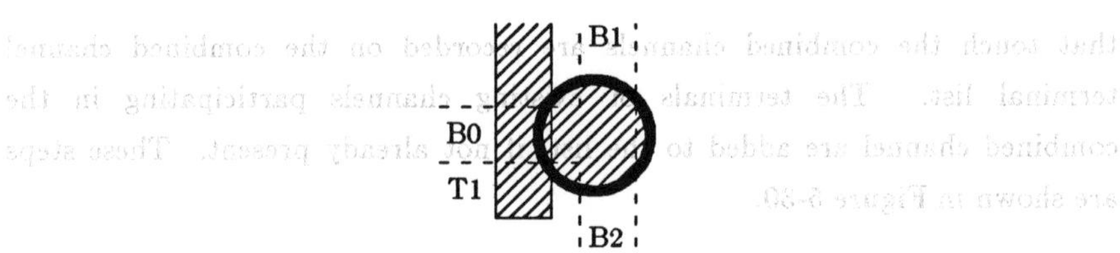

- The combined channel has less than two terminals, and does not touch existing channels. A capacitor is formed, but ignored, since it is not a DC topology change.

- The combined channel has two terminals and also touches an existing channel. If the combined channel is a different shape than the existing channel, then a device size change has occurred. If the combined channel is the same shape as the existing channel, then no size change has occurred. In either case, no DC topology change occurs, so no fault is reported.

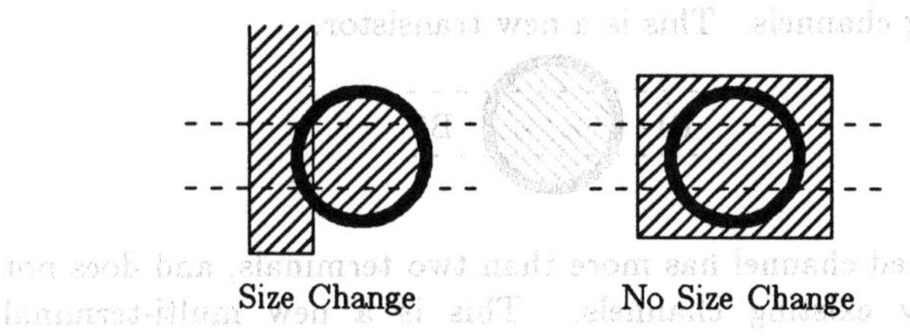

- Otherwise no transistor is formed.

The above rules depend on the assumption that only devices with one source and one drain terminal are permitted in the original circuit. By using the *change* in transistor terminal count, rather than absolute number of terminals, it is possible to deal with multi-terminal transistors in the original circuit. However this algorithm would be more complex. Since multi-terminal transistors are not normally used, we disallow them, and use a simpler fault analysis procedure.

5.4.7. New Active Device

A new active device occurs when an extra active defect spans a poly line, forming a channel and a source and drain, as was shown in Figure 3-7. If a new device occurs, the analysis procedure reports a new device including its gate and terminal connections.

New active devices are detected by first subtracting from the defect those gate (poly) polygons that do not intersect buried contact, creating a masked defect. This is shown in Figure 5-31.

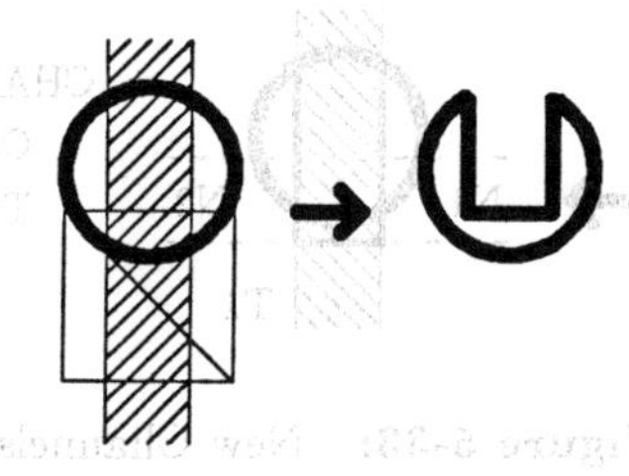

Figure 5-31: Masked Defect

The masked regions form poly-active contacts rather than transistors. If the masked defect does not have at least two contours, then either no new channel was created, so no new device can exist, or the new device has only one terminal, and so is a capacitor, and so is discarded, as is the case in Figure 5-31. If the masked defect has at least two contours, it is decomposed into separate masked defects. The net numbers of active polygons that touch the defect pieces are recorded as shown in Figure 5-32. Part of this work is shared with the short circuit fault analysis procedure.

The gate (poly) polygons with buried contact subtracted are intersected with the original defect, creating new channel regions. These new channel regions are unioned with all existing channels that they touch, forming combined channel regions. The existing channels used in each combined channel are recorded. The terminals of each existing channels participating in a combined channel are added to the combined channel terminal list. This is shown in Figure 5-33.

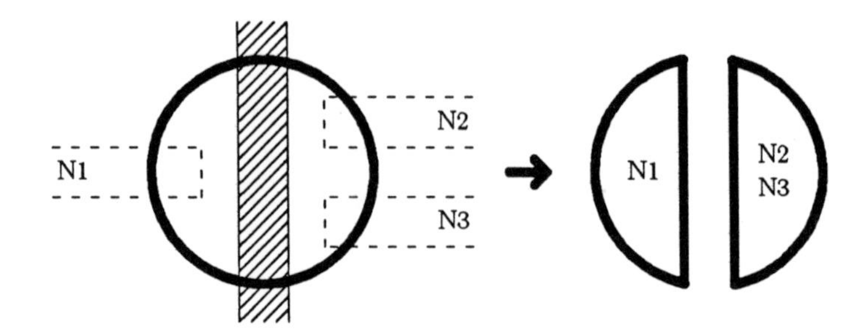

Figure 5-32: Labeled and Decomposed Defect

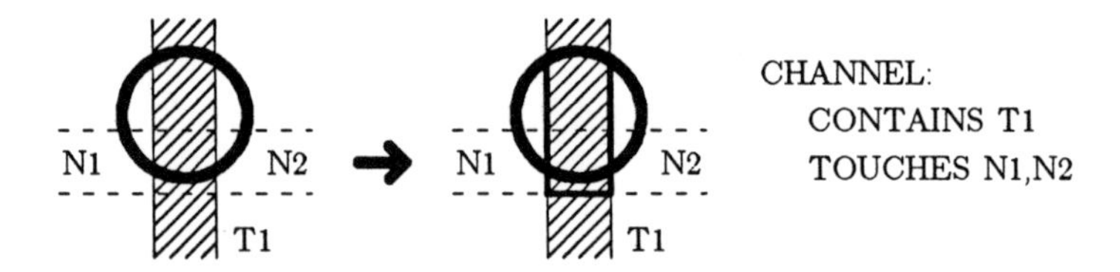

Figure 5-33: New Channels

Some pieces of the masked defect touch a channel, but do not touch other active polygons, so do not have a net number assigned to them. A new net number is created, called a *generated* net number, and the terminal is added to the channel terminal list. These new terminals will either be floating, and so discarded, or they will connect to another new transistor. The latter situation is shown in Figure 5-34.

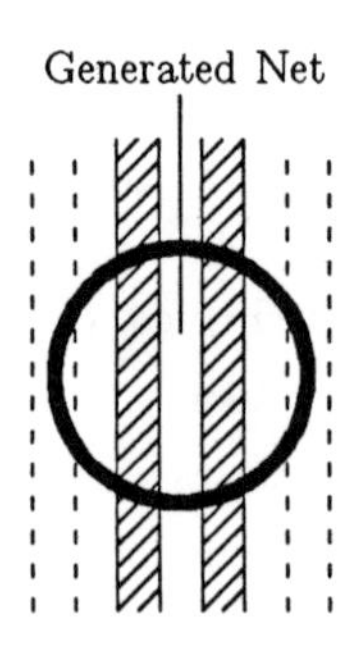

Figure 5-34: Two New Active Devices

This configuration rarely happens since it requires several poly and active lines close together along with a large extra active material defect. If poly line widths and spacings are three microns, and poly-active spacing is two microns, then the defect must be at least 13 microns wide to cause two new devices.

A new active device is created in the following cases:

- The combined channel has two terminals and does not touch any existing channels. This is a new transistor.

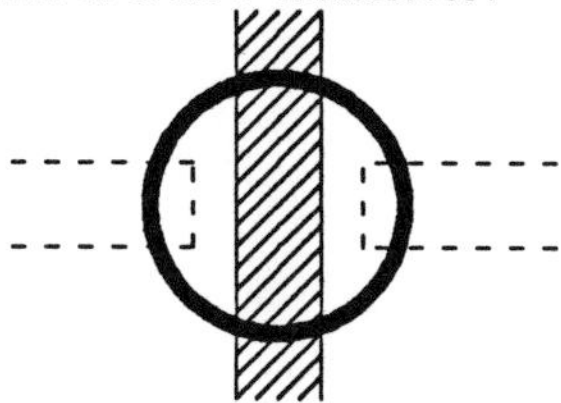

- The combined channel has more than two terminals, and does not touch any existing channels. This is a new multi-terminal transistor.

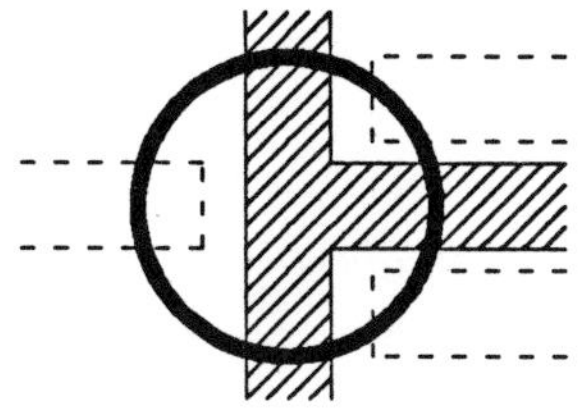

- The combined channel has more than two terminals and touches existing channels. This is a multi-terminal transistor converted from existing transistors.

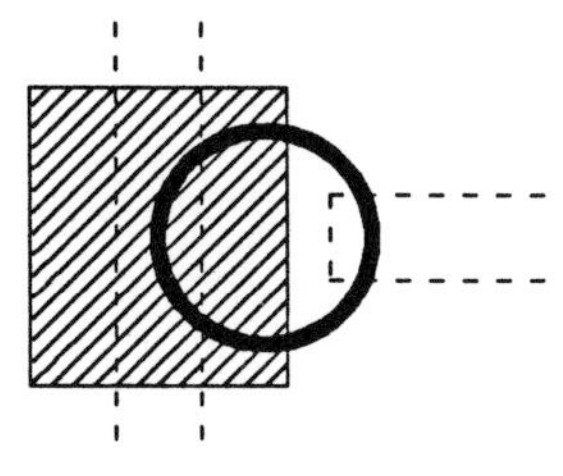

- The combined channel has two terminals and touches existing channels. If the combined channel polygon is not equal to the original channel polygon, then a size change has occurred. Otherwise no size changed has occurred. In either case, no DC topology change has occurred, so no fault is reported.

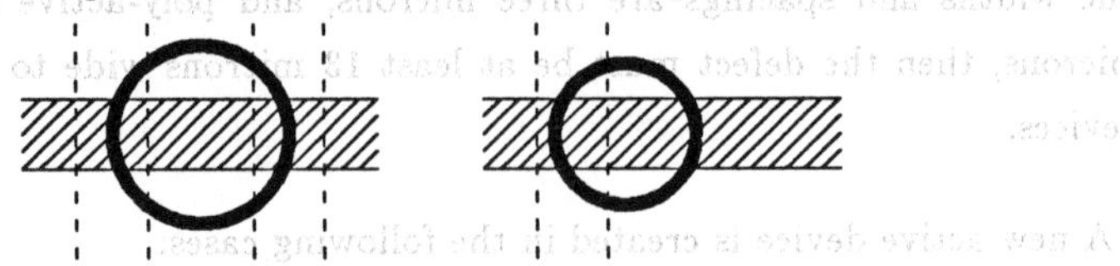

- Otherwise no transistor is formed.

As was the case for the new gate device procedure, the new active device analysis procedure requires that there be no multi-terminal devices in the original circuit description. Again, by considering the change in the number of transistor terminals rather than the absolute number, it is possible to allow multi-terminal devices. However since these devices do not ordinarily occur, and their analysis complicates the algorithm, they are not permitted.

When examining each combined channel and its terminals, if a device is created with a terminal that has a generated net number, then the device is put on a list. After all new devices have been created, those with generated net numbers are checked to see that at least two new devices share a terminal with a generated net number, as is the case in Figure 5-34. If this is not the case, then the terminal with the generated net number is floating, and must be discarded. This reduces the terminal count for the device. The above case analysis must be done again to determine whether the device qualifies as a new active device fault, and what type.

When an active polygon edge is coincident with a gate edge, then a new active device fault may be missed. These situations are shown in Figure 5-35. In both cases, the masked defect has only one contour, so no new device is reported. These geometry configurations are normally disallowed, so we do not attempt to analyze them.

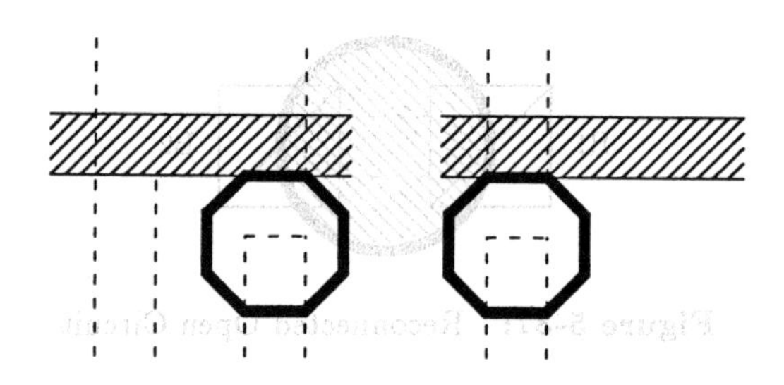

Figure 5-35: Missed New Active Device Faults

5.5. Fault Combination and Filtering

Some defects may cause several different faults to occur. In some cases, these faults must be combined into a simpler fault, or discarded altogether. For example, an extra poly defect may block a metal-active via. This causes an open between the metal and active, and causes the metal to short to the extra poly. At the same time, the extra poly may also short to the active through a buried contact, as shown in Figure 5-36.

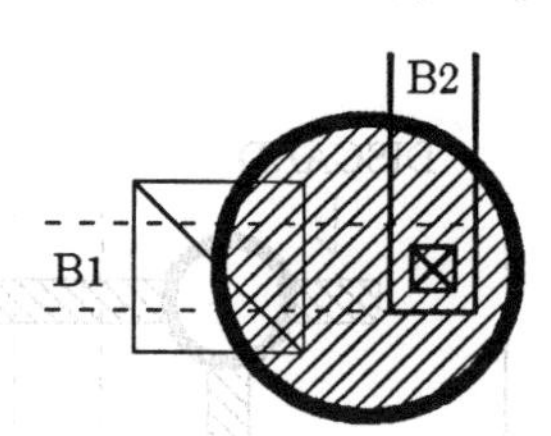

Figure 5-36: Reconnected Blocked Via

The path from the metal via through the poly to the buried contact recompletes the circuit, so no fault has occurred. Similarly, an extra poly defect can break an active line in the process of forming a new device. However the poly may connect to the active on each side of the break through buried contacts, reconnecting the active net, as shown in Figure 5-37.

Both of the above cases use the information gathered by the branch traversal in the open circuit fault analysis procedure. The traversal notes

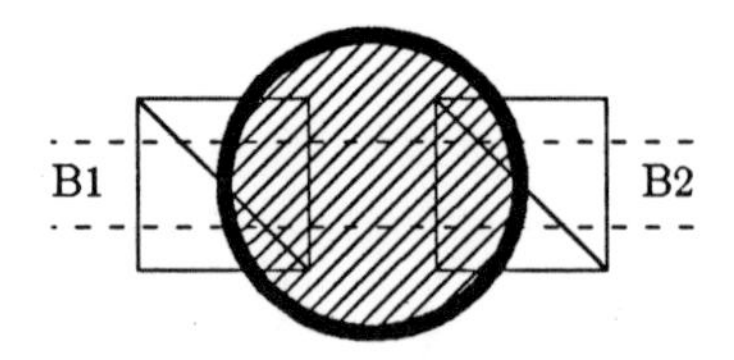

Figure 5-37: Reconnected Open Circuit

that both branch 1 and 2 are connected to the via, and so are reconnected.

Some faults require information from several of the fault analysis procedures. Short circuits are used to connect the gate of a new gate device. The branches of an open circuit are connected to the terminals of a new gate device.

Some faults are discarded because they do not cause DC changes to the circuit topology. An open circuit results in several branches of the broken nets. Those branches that do not connect to transistors or cell boundaries cannot change the circuit topology, and so are discarded, as shown in Figure 5-38.

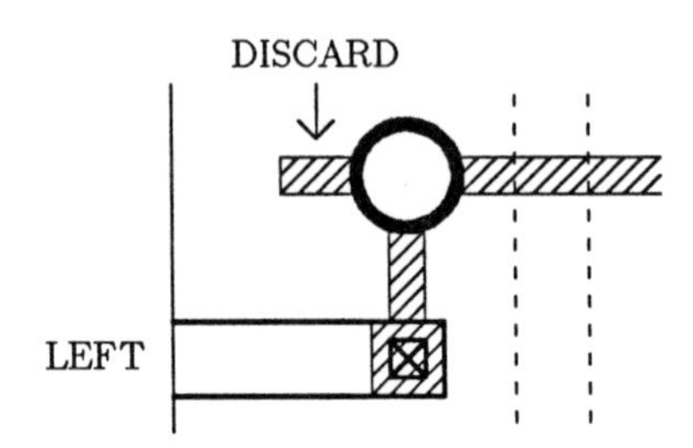

Figure 5-38: Discarded Open Branch

If there is only one remaining branch, then the open circuit is discarded. Since the branches of an open circuit can form the terminals of a new gate device, discarding branches may change the terminal count of a new gate device. If there are not at least two source/drain terminals in a new device, or three in a device involving an existing transistor, then the new gate device must be discarded.

The fault type of new vias that connect two nets is converted to short circuit. Those new vias that short a transistor gate to channel are left as new vias. The reason for doing so is that a gate to channel short does not have a well-defined electrical meaning. The fault as a separate type, so that the application program can interpret its meaning.

Finally the fault combination phase sorts each circuit fault and the list of faults into a canonical order. This makes comparison between two fault lists easier in later processing steps.

Chapter 6
VLASIC Implementation

This chapter describes the VLASIC implementation. A functional description of the simulator was provided in Chapters 1, 3, 4, and 5. This chapter describes how these functions are implemented, with particular emphasis on the polygon package. We then describe example VLASIC simulations, and discuss VLASIC's performance and accuracy.

The VLASIC implementation is divided into modules as shown in Figure 6-1. The *process tables* contain process descriptions and defect models used in the data preprocessing, fault analysis, and fault combination and filtering modules. The *polygon package* provides the basic polygon operations used for all geometry processing. The *preprocessing phase* reads in the extracted layout, and builds the global layout data structures. The *distribution module* reads in the process parameters and wafer maps to initialize the hierarchical random number generators used to generate and place defects on the layout. The *control loop* generates chip samples, calls the fault combination and filtering module, and then stores the resulting faults. The *fault summary* module stores and summarizes the chip fault lists. The *fault combination and filtering* module calls the fault analysis procedures, and then combines their results and filters out non-DC topology changes. The *fault analysis* module contains fault analysis procedures for the basic circuit fault types. The following sections describe the implementation of these seven major modules.

The VLASIC simulator contains approximately 16,000 lines of C code implemented on a VAX running Berkeley UNIX. Another 5500 lines of code are required to interface to the Magic layout system.

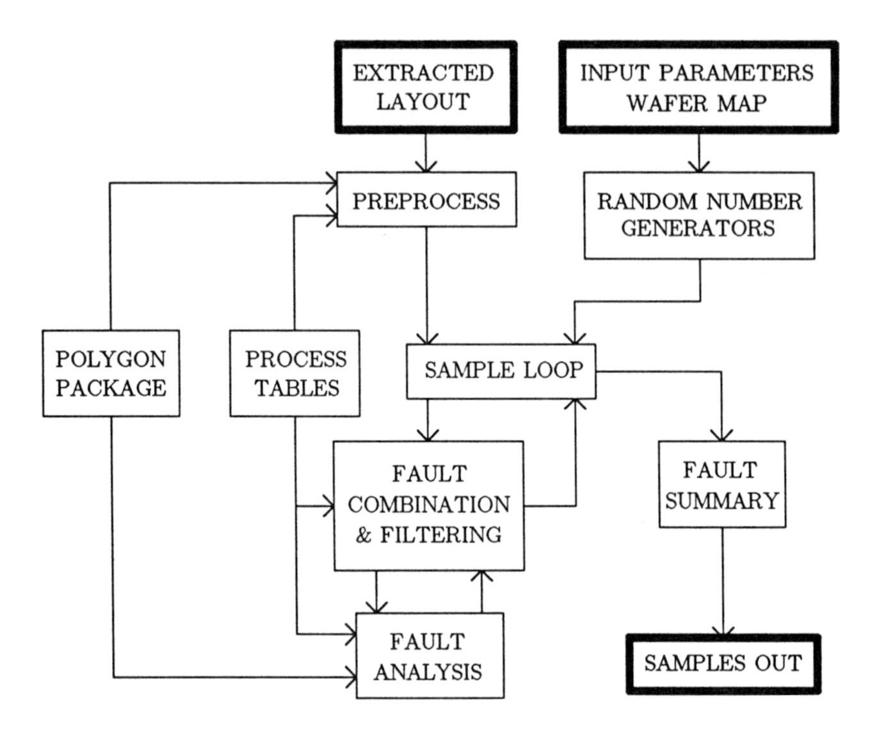

Figure 6-1: VLASIC Implementation

6.1. Preprocessing and Data Input

The chip layout must first be extracted in order to label all geometry with net numbers and locate all transistor channels. For a CIF (Caltech Intermediate Form) layout description, the ACE circuit extractor [Gupta 81, Gupta 83a] is used, generating a wirelist [Frank 81]. This wirelist contains a list of geometry for each net and each channel. The wirelist can be read directly into VLASIC. Since the wirelist takes about 800 bytes per transistor, it can also be packed into a binary format that is three times smaller. This packed format can also be read directly into VLASIC. For layouts done using the Magic layout system [Ousterhout 84a, Ousterhout 85], Magic's integral circuit extractor is used to extract the circuit. This extracted circuit does not contain net or channel geometry. This geometry is attached by combining it with a Magic-generated CIF file, resulting in a netlist. This netlist is then converted to the packed format for use by VLASIC. Since VLASIC uses only fully-instantiated geometry, the Magic files must be flattened before extraction and CIF generation. The algorithm used to attach the CIF geometry to the extracted nets is a modified version of the algorithm used for following broken nets in the open circuit fault analysis procedure. This algorithm is optimized for small problems, and its quadratic behavior currently limits us to small Magic-generated layouts.

The data input phase also reads in a wafer map for use by the random number generators. The wafer map is stored as a file of rectangles located relative to the wafer center.

The process parameters are stored in a process file, or can be provided on the command line. These parameters specify the defect density, between-lot and between-wafer clustering coefficients, radial distribution, number of wafers per lot, diameter of peak frequency, minimum resolvable width and spacing, and minimum interesting defect diameter.

The layout is stored in an array of bins for efficient processing. These bins are used to obtain linear expected time geometry manipulation algorithms [Bentley 80]. Each rectangle is inserted into a list for each bin it touches. Ideally the bin array is chosen to be $\sqrt{N}$ by $\sqrt{N}$ for N rectangles, so that there is approximately one rectangle per bin. Due to memory limitations, we have chosen to make bins 32 microns square, so there are often 3-4 rectangles per bin. A separate array of bins is used for each layer.

The layers are then merged, intersected, and subtracted to generate layer combinations as specified by the process tables. These layer combinations are used to speed up the fault analysis procedures. For example, a special layer specifying metal1-metal2 vias is created by intersecting the metal1, metal2, and second-level via layers. Only those layers that have small disjoint (unconnected) polygons, such as the contact layers, and overlap layers, are merged. Otherwise large merged polygons would be created. As discussed below, large polygons are undesirable for reasons of space and time.

The polygon data structures include information to aid in fault analysis. Each polygon is labeled with its layer, and its net numbers. A polygon representing a layer combination can have several net numbers. In the processes under consideration, only one or two net numbers can occur, since there are at most two parallel conductors in any layer combination. If the polygon forms part of a transistor channel, it also has a pointer to a structure describing the transistor. Each polygon also has a flag indicating whether it should be saved, or can be deleted after some intermediate operation. Polygons also have a flag indicating whether they have already been used during the current operation. Since the same polygon can reside in several bins, algorithms that scan the lists of polygon in several bins could encounter the same polygon more than once. This flag is also used during net traversal in the open circuit fault analysis procedure.

6.2. Process Tables

The process tables describe the details of the process relevant to VLASIC, such as which layers are interconnected by a via, how to generate the layer combinations, how to combine the nets of each layer when generating a layer combination, whether to merge, intersect, or subtract layers, and whether to delete intermediate results. The tables also specify what circuit faults might be caused by a particular defect type. They specify what layers interact with a defect for a particular fault type. They specify whether a defect is positive (additional conductor) or negative material, and on what layer a positive defect lies (e.g. extra metal lies on the metal layer). A table specifies what layer combinations contain source/drain layers and what combinations contain gate (poly) layers. A table specifies what layers mask what defects for what fault types. For example, the active-buried contact combination masks extra polysilicon for open circuits. Another table specifies what layer combinations become completed vias when adding a positive non-via defect. The table also specifies what positive layer is needed to complete an incomplete via, what layers can possibly participate in an incomplete via, and what layer combinations are vias. These via tables are used by the short and open circuit fault analysis procedures.

6.3. Random Number Generators

The defect random number generators must generate Poisson, negative binomial, and uniform integer deviates. For mean $\lambda \leq 12$, the Poisson deviate is generated directly from the exponential PDF [Knuth 81]. This algorithm has a running time linear in λ. The mean can be quite large when representing the number of defects in a zone of a wafer. Therefore for larger λ, the method of [Ahrens 80] is used. This algorithm has essentially constant running time for large λ. For $\lambda \leq 10$, negative binomial deviates are generated directly from the PDF. The running time grows linearly with λ, and inversely with clustering coefficient α. Since the negative binomial is used to determine the number of defects in a lot, λ can be quite large. For large λ, the deviate is calculated using the method of Le'ger. A sample X is taken from a standard gamma distribution with parameter α. A sample K is generated from a Poisson distribution with mean $X\lambda/\alpha$, and then output [Knuth 81]. A fast three-layer gamma distribution is implemented using the techniques described in [Ahrens 74]. This gamma distribution in turn uses a normal distribution implemented as a three-layer distribution [Nassif 84b]. The uniform deviates required for all of these algorithms are generated with a linear feedback shift register with a period of 2^{36}. This period is adequate since no more than 50-100 deviates are needed per chip sample simulated. The linear feedback register algorithm is fast, accounting for only 0.1% of the execution time.

The defect size distribution is implemented directly from the PDF. We currently use the untruncated size distribution. A truncated distribution has the advantage that fewer defects need to be placed, however there is only a small performance improvement, as discussed later in the chapter.

The defects are placed uniformly within a die by generating a random (x,y) location. The location is specified as a pair of integers representing centimicrons rather than as real numbers, due to precision problems. Rounding to centimicrons does not introduce significant errors into the simulation since polygons and defects are many centimicrons in size.

6.4. Control

The control loop parses the inputs, does the preprocessing, and then goes into a sample loop using the random number generators to create and place defects on the layout. It then calls the fault combination and filtering module. The resulting faults are stored away in a summary data structure. When the simulation is complete, this summary is printed out.

6.5. Fault Analysis

The fault analysis module is divided into the seven fault analysis procedures described in Chapter 5 plus support routines. The fault analysis algorithms have already been described. The support routines provide common functions for masking defects, and merging polygons in a list or a set of bins.

6.6. Fault Combination and Filtering

The main control loop calls the fault combination and filtering module after a defect has been placed, specifying the defect diameter, location, and type. Since defect diameters are rounded to centimicrons, and are bounded by a maximum diameter, there are only a finite number of defect polygons. These defect polygons are created as necessary and stored in a cache. Since the maximum defect diameter is currently 18 microns, there can be 1800 possible defects. However since the defect size distribution is highly skewed, typically only half of all possible defect diameters are generated when simulating 25,000 defect placements. These defects are stored at the origin, and moved to their location on the chip using the MOVEPOLYGON function. When fault analysis is complete, they are returned to the origin.

After the defect polygon has been selected, the defect neighborhood must be determined. Given the defect location and diameter, the indices of those layout bins that touch the defect are calculated. The combination module then calls all those fault analysis procedures appropriate for the defect type. In the case of short circuits and new active devices, some work is shared masking an extra active defect and connecting it to surrounding active polygons. This work is done once and then passed to the two fault analysis procedures. The fault groups returned by each analysis procedure are stored on a chip fault list.

After fault analysis is complete, the fault lists are first sorted into a canonical order. Net numbers and terminals are sorted into ascending order within each fault. Faults are then sorted in lexicographic order. The fault sorting is used to simplify the comparison of different faults during fault combination and fault summarization.

Open circuit faults caused by missing material are then processed to discard those that only have one branch touching transistors or bin boundaries. Open circuits caused by extra material are then processed. The open circuit procedure returned a list of all those broken net branches that connect to the defect. Those branches that do not touch transistors or bin boundaries and that are not connected to the defect are discarded. If a branch does connect to the defect, then it can form the terminal of a new gate device. Finally all those branches that are electrically connected together through the defect are combined together.

The new gate devices are processed to connect the gate up to any shorts, and to connect the source/drain terminals to open branches. The branch labels used in the open circuit and new gate device are the same to facilitate this connection. Those terminals that do not connect to an open branch are discarded. The reason an open branch might not exist is that it has previously been discarded as floating. When transistor terminals are discarded, the new gate device fault must be reexamined to determine whether it should be modified or discarded. If a new transistor has zero or one terminals, it is discarded. If a new multi-terminal device has only two terminals, it is converted into a new transistor. If a multi-terminal device involving existing transistors only has two terminals, it is discarded. One-net shorts are discarded since they have been connected to the new gate devices, and are not valid shorts by themselves.

Finally those new vias faults that are junction leakage shorts, or intermediate oxide shorts are converted into short circuits. This is a notational convenience for post-processors.

6.7. Fault Summary

The fault lists are stored in a summary database. When simulation is complete, they are printed out. The fault combination module returns the fault group caused by a defect. Each fault in the group is compared with those that have already been generated. The fault is saved only if it is unique. This saves a large amount of space since some faults occur many times. The comparison process ignores the defect location and diameter, otherwise almost every fault would be unique. The comparison process only uses the fault type, the type of defect causing the fault, and the details of the fault.

When all the defects on a chip sample have been processed, the chip fault list is compared with those that have already occurred. If the list is unique, it is stored in the database. If the list has already occurred, the frequency count for that list is incremented. Again this saves a large amount of space, since as will be seen in the examples, the fault distribution is highly skewed.

6.8. Polygon Package

VLASIC represents layout geometry as a set of polygons. The fault analysis techniques described in Chapter 5 use polygon operations such as UNION, DIFFERENCE (subtraction), and INTERSECTION in order to detect whether a defect placed on the layout causes a circuit fault to occur. These polygon operations are implemented by a polygon package. This polygon package is based on work by Faust [Faust 80a, Faust 80b, Faust 81a, Faust 81b, Faust 81c]. Faust's work is based on work by Barton and Buchanan [Barton 79, Barton 80b], which in turn is based on work by Sutherland [Sutherland 78]. Our package has been extensively modified and extended to achieve correctness, improve performance, use finite precision, and to add special functions required for fault analysis. In this section we describe how polygons are represented, the basic polygon operations implemented by the package, the theory behind the polygon operations, special functions provided for use by the fault analysis procedures, and finally performance.

6.8.1. Polygon Representation

The polygon package represents polygons as a list of contours. Each contour consists of a list of directed edges, as shown in Figure 6-2.

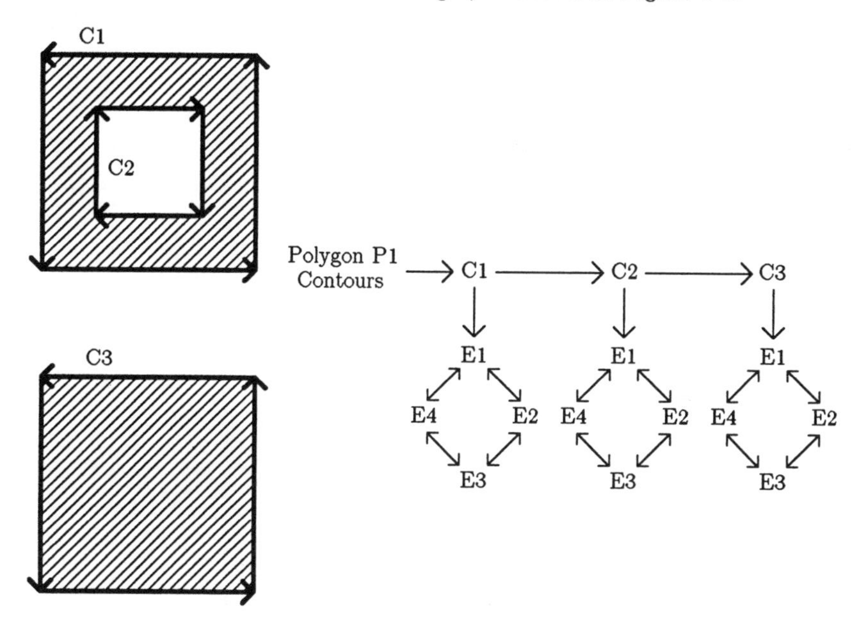

Figure 6-2: Polygon Contours

The interior of a polygon is to the left of an edge when going from tail to head, so a counterclockwise contour such as C3 represents an enclosed space, while a clockwise contour such as C2 represents a hole. Counterclockwise and clockwise contours are also referred to as positive and negative contours respectively. Note that disjoint regions are allowed, such as the regions formed by C1-C2 and C3. Edges must be straight lines. This means that we must approximate a circular extra or missing material defect by an octagon, as noted in Chapter 3. This also has implications for expanding and shrinking polygons as discussed below.

Contours are stored on a singly-linked list. Each contour is a doubly-linked circular list of edges with an edge count. The polygon also has a bounding rectangle (box).

In order to save space, rectangular polygons are compressed into rectangles represented as a lower-left and upper-right point. These rectangles are expanded into temporary polygons whenever polygon operations are done on them.

6.8.2. Basic Polygon Operations

The basic operations on polygons are UNION, INTERSECTION, DIFFERENCE, and EXPAND. These operations are illustrated in Figure 6-3.

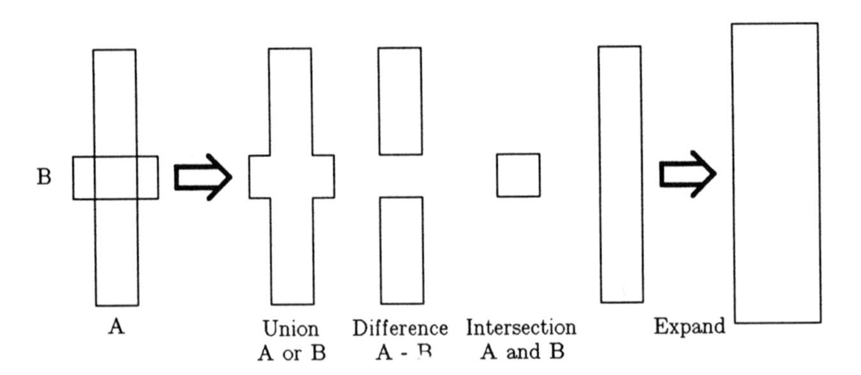

Figure 6-3: Basic Polygon Operations

The UNION and INTERSECTION operators generate the disjunction and conjunction of two polygons. The DIFFERENCE operator subtracts one polygon from another. The EXPAND operator either inflates or deflates a polygon depending on whether expansion is by a positive or negative amount. The ISECT (touch) operator reports whether two polygons intersect, but does not generate the actual intersection polygon.

The EXPAND operator uses orthogonal geometry. A polygon is expanded by +5 units by increasing the length of each edge by 5 at each endpoint, and moving the edge 5 to the right, when looking from tail to head. This is illustrated in Figure 6-4.

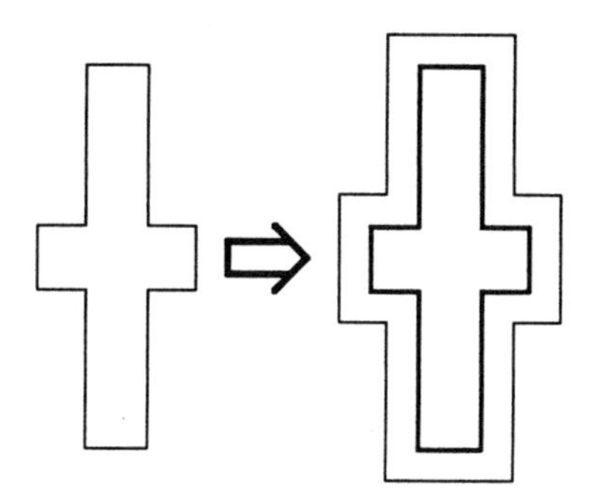

Figure 6-4: Orthogonal Expansion

In Euclidean geometry, the EXPAND operator creates a new contour with all points moved +5 out from the old contour. At corners this means that arcs are created, as shown in Figure 6-5.

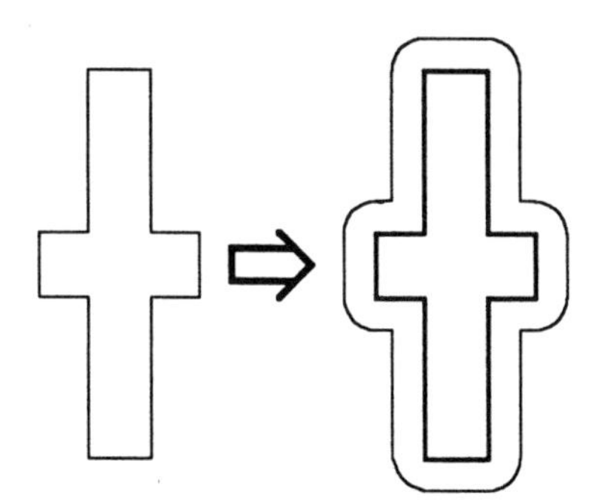

Figure 6-5: Euclidean Expansion

Euclidean and orthogonal deflation are equivalent. Euclidean geometry is required for some applications. For example, in order to determine whether two polygons are at least 10 units apart, each polygon is expanded by +5, and then an ISECT test is performed. If there is no intersection, then the polygons are at least 10 units apart. However if inflated polygons are to be used in later operations, then all polygon operators must handle contours composed of arcs and straight edges. As was noted in Chapter 3, such polygon packages are very slow.

In polygons with holes, expansion can cause the edges on opposite sides of a hole to pass each other, converting the contour from clockwise to counterclockwise, or from a hole to a positive contour. This results in an extraneous contour inside the polygon, as shown in Figure 6-6.

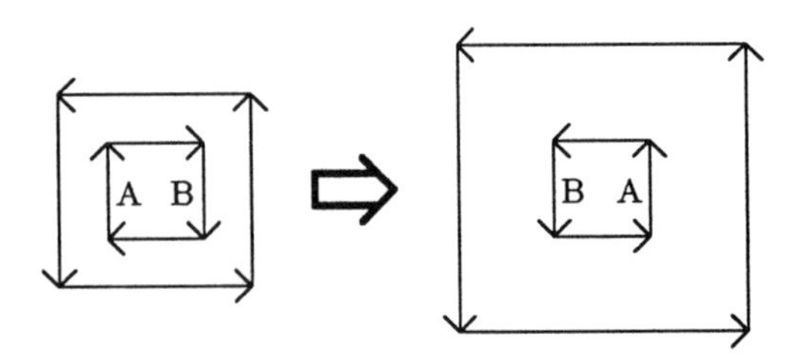

Figure 6-6: Disappearing Hole

This change of direction is detected by noticing that the edges of the contour have changed direction by 180 degrees. These contours are then discarded.

When a polygon is expanded by -5, each edge endpoint is moved in by 5. This can potentially cause the head and tail to exchange places. The edge is also moved 5 to the left when looking from the original tail to head. When an edge moves left, it can pass an edge on the opposite side of the polygon moving right, exchanging places. Because the edge endpoints have also swapped places, the contour is still counterclockwise, as shown in Figure 6-7.

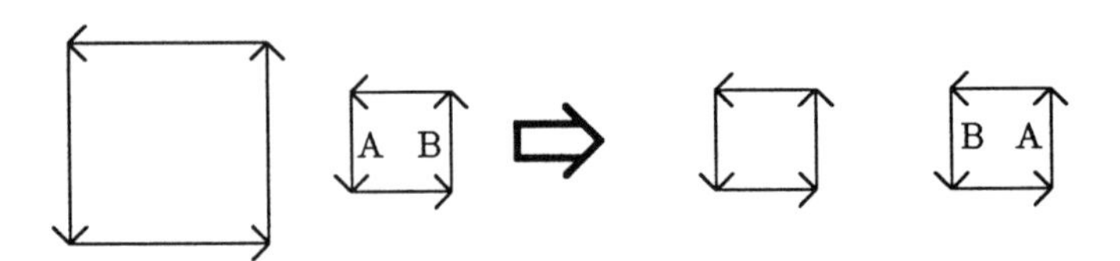

Figure 6-7: Disappearing Polygon

Since the contour is counterclockwise, it represents a polygon. The correct result should be that the polygon shrinks to nothing and disappears. Again this situation is detected by noticing that the edges have flipped by 180 degrees. Other polygon packages have not always handled this situation correctly, particularly in the case of concave polygons, or polygons with holes.

6.8.3. Polygon Package Algorithms

The UNION, INTERSECTION, and DIFFERENCE operators use the same basic algorithm. The two polygons are examined to determine all those points where an edge of one polygon intersects the edge of the other polygon. The rules for whether an edge intersects another are determined by the operator. The ISECT operator uses this edge intersection list to determine whether a polygon intersection has occurred. For each of the other operators, an intersection point is chosen and the contours of the output polygon are determined by traversing along the input polygon edges from intersection point to intersection point, changing paths according to the operator rules. At each intersection point there are two possible edges to follow. As part of the selection process, the midpoint between the intersection point and the next edge endpoint or intersection point is determined. Winding (wrap) numbers are calculated for these midpoints using the method of Guibas et al. [Guibas 83], as shown in Figure 6-8.

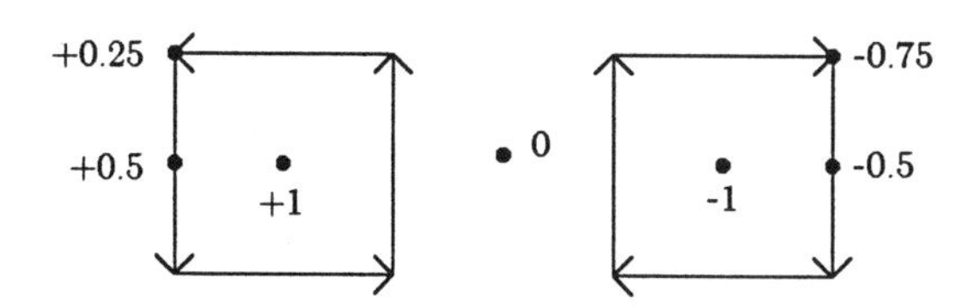

Figure 6-8: Winding Number Calculations

A point inside a positive contour has a winding number of +1. A point inside a negative contour has a winding number of -1. A point outside a polygon has a winding number of 0. For a point on an edge, but not at an endpoint, the winding number is +0.5 for a positive contour, and -0.5 for a negative contour. At a corner, the winding number is the fraction of 360 degrees that the angle to the left of the contour makes when following edges from tail to head. A 90 degree angle on a positive contour has a +0.25 winding number. A 90 degree angle on a negative contour has a -0.75 winding number. The corner calculation is somewhat expensive, but it occurs only 2-6% of the time in yield simulations.

For the UNION operator, the edge we want to traverse (put in the output polygon) is the one whose midpoint lies on the boundary of either or both polygons, as shown in Figure 6-9.

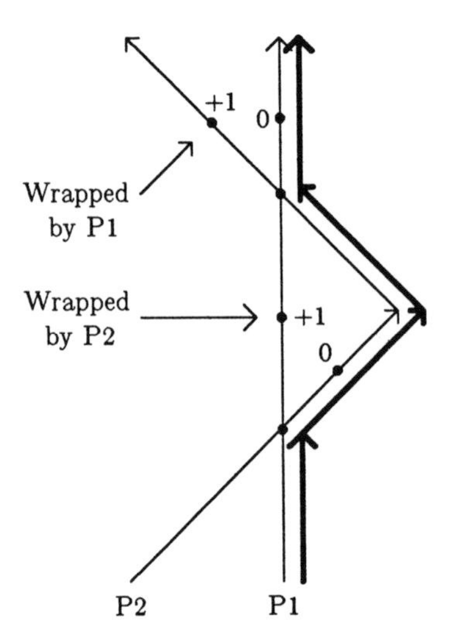

Figure 6-9: Following Polygon Boundaries in Union

A major improvement of our polygon package over existing packages is the use of this improved winding number calculation. This calculation eliminates the need to worry about many special cases. These special cases tended to be the source of many bugs, particularly when dealing with non-rectilinear polygons, such as octagonal extra and missing material defects.

Our polygon package originally used integer arithmetic. This caused roundoff errors in the edge intersection calculation. Roundoff errors still occurred when double-precision arithmetic was used. These errors mean that if $C = A \cap B$, then $D = A - C$ is not the same as $D = A - B$. This led to the creation of polygons containing a crevice or sliver as shown in Figure 6-10.

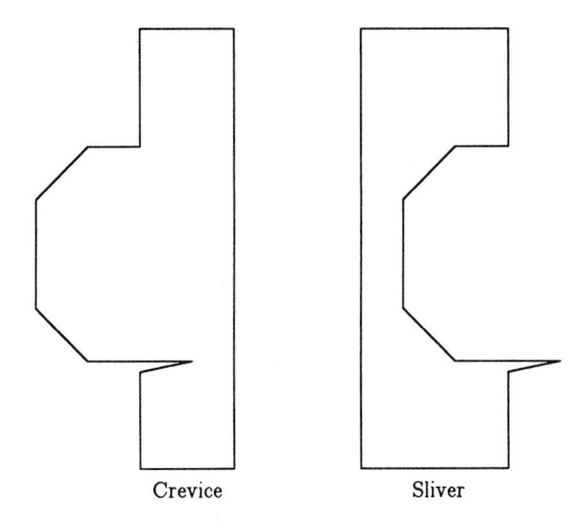

Figure 6-10: Polygon Pathologies Due to Roundoff

Since design rules are typically a large fraction of a micron or more, accuracy can be maintained at the centimicron level without introducing significant errors in the yield simulation. This allows the use of a finite precision polygon package, which avoids the precision problems described above [Milenkovic 85]. All points are rounded to the nearest grid unit. However in these finite precision systems, it is possible for polygon edges to be fractured into several segments to force intersection points to grid points. This is referred to as *edge cracking*. This cracking can accumulate over several operations, resulting in severe performance degradation. Given that our fault analysis operations involve arbitrarily placed polygons with small angles, we felt that cracking was a real possibility. Hence a finite precision polygon package was not used. In systems that only deal with rectilinear polygons, edge cracking is not a problem.

The solution that we adopt is to say that if a point B is within some minimum distance L from the line segment AC formed by two other points A and C, then the points A, B, and C are colinear, as shown in Figure 6-11.

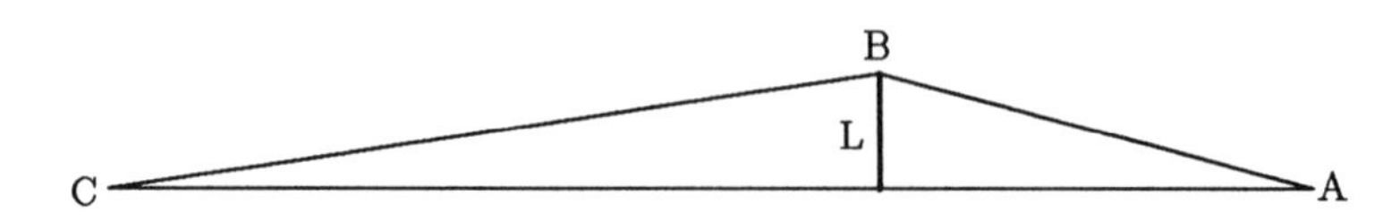

Figure 6-11: Colinear Points Determination

Unlike a finite precision polygon package, segment AC is not broken to insert point B, so edge cracking does not occur. However like a finite precision package, points that are close together are treated as equivalent.

In most cases, slivers can be eliminated by a small deflation and inflation, while crevices can be eliminated by inflation and deflation. However this has undesirable properties. For example, when detecting whether two lines are shorted together, if the lines are ϵ apart, they are not electrically connected. However if the edges are slightly inflated and deflated, then they will be joined, which is incorrect.

6.8.4. Special Functions

A standard polygon package only provides boolean operations on sets of polygons or individual polygons. The fault analysis procedures described in Chapter 5 require additional polygon operations to conveniently perform fault analysis. These special functions are:

- MOVEPOLYGON(X,Y,P) - translate polygon P by (X,Y). This function is used to move defect octagons from place to place on the layout. This is much faster than generating a new polygon for each defect placement. Defects can be reused since their diameter is an integer number of centimicrons, so there are only a finite number of possible defects.

- NUMCONTOURS(P) - return the number of contours in P. This function is used by most of the fault analysis procedures to determine whether the defect polygon has spanned a line.
- ISRECTANGLE(P) - return true if P is a rectangle. Most layout polygons are rectangles. Rectangular polygons can be stored in less space and more rapidly processed than general polygons.
- ADJACENT(P1,P2) - return true if P1 and P2 share an edge. Edge adjacency is used during coincidence tests in open circuit fault analysis.
- BURSTPOLYGON(P) - break apart the disjoint contours of P into several polygons, as shown in Figure 5-12. This function is used in detecting open circuits, shorted and open devices, and new gate devices.
- ENCLOSES(X,Y,P) - return true if point (X,Y) lies in the interior or on the edge of P. This function provides a rapid test for whether a new via fault has occurred or not.
- EQUALS(P1,P2) - return true if P1 and P2 are equivalent. This function is used to determine whether a new active device fault has occurred, or whether a device size has changed.

6.8.5. Polygon Package Performance

The performance of the polygon package is critical because the execution time of VLASIC is dominated by polygon operations during the fault analysis phase.

The time to perform a union, intersection, or difference on an n and an m edge polygon is $O(mn)$. This results from the fact that every edge in one polygon is compared to every edge in the other polygon to determine whether an intersection occurs. In the worst case, $O(mn)$ time is the the minimum necessary since for two star-shaped polygons, there can be nm intersection points [Lee 84]. The winding number calculation (which implements the ENCLOSES operation) on an n edge polygon takes $O(n)$ time. This is the minimum necessary in the worst case, which is a spiral-shaped polygon.

It is possible to detect whether a polygon intersection has occurred (ISECT) in $O(N \lg N)$ time where $N = m + n$ by sorting edges. If the number of intersection points is roughly constant, then polygon operations can be done in $N \lg N$ time. This is done by using a set of hierarchical bounding boxes for edges and groups of edges [Barton 80b]. If the polygons are convex, then binary operations can be done in $O(m + n)$ time using the plane sweep method [Lee 84]. For convex polygons, the winding number can be calculated in $\lg n$ time by searching to determine which edges lie to the right of the point. These algorithms typically use more space than the simple algorithm. As discussed below, we cannot afford this space, so the faster algorithms are not used.

For polygons with 4-10 edges, the time penalty for our $O(mn)$ time algorithm is small because first the bounding boxes of contours are compared, and then the bounding boxes of edges. Only when edge bounding boxes overlap is an expensive edge intersection test done. For example, an ISECT between a disjoint rectangle and octagon takes 0.24 msec on a VAX-11/785 while an ISECT between an intersecting rectangle and octagon takes 1.97 msec. The expense in a boolean operation is primarily in calculating the winding numbers and determining whether an intersection has occurred using finite precision techniques. However the penalty is large enough that it is not possible to merge all the polygons on a layer and obtain adequate performance. The circuit extractors and layout systems used to generate VLASIC input break the layout into relatively small rectangles. As discussed in Chapter 5, we only merge these polygons in the neighborhood of a defect when performing the fault analysis.

The polygon data structures require 173 bytes for a rectangular polygon. Rectangles are detected and compressed into 29 bytes, and expanded into temporary polygons before use. This compression and expansion adds 80 microseconds to a polygon operation, which is up to a 40% increase in time. Even with rectangle compression, virtual memory space is the primarily limitation on the size of the problems that can be handled by VLASIC. About 1750 bytes per transistor is required, so that a 3000 transistor problem takes 5 megabytes. As discussed in Chapter 9, a hierarchical version of VLASIC will greatly reduce space requirements, and allow the use of faster polygon algorithms.

6.9. VLASIC Examples

6.9.1. Dynamic RAM Cell Example

The VLASIC simulator was used to simulate the fabrication of a "chip" containing a single three-transistor dynamic RAM (DRAM) cell. The layout of the cell is shown in Figure 6-12. The chip is placed at 100 locations on the wafer as shown in Figure 6-13.

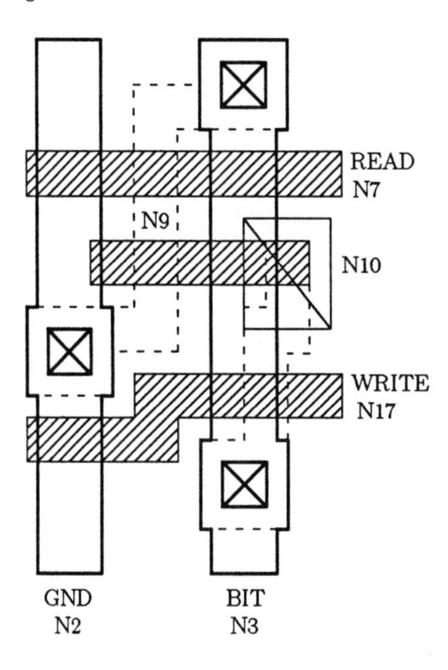

Figure 6-12: Three Transistor DRAM Cell

The synthetic process conditions used in the simulation are given in Table 6-1. A very small circuit was chosen as an example in order to limit the number of unique circuit faults generated by the simulation. The very high defect densities are chosen in order to obtain an average of 2.5 defects per

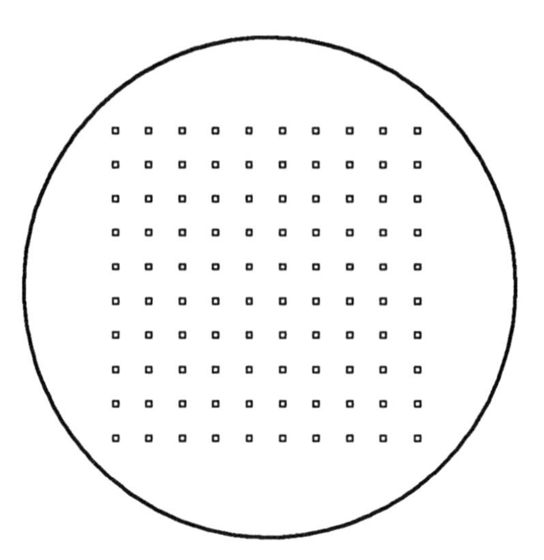

Figure 6-13: Wafer Map

Defect Density	
Extra/Missing Metal	20,000/cm2
Extra/Missing Poly	20,000/cm2
Extra/Missing Active	20,000/cm2
First-Level Pinholes	20,000/cm2
Gate Oxide Pinholes	20,000/cm2
Design Rules	
Metal Width/Space	6u
Metal Contact Width	4u
Poly Width/Space	4u
Active Width/Space	4u
Poly/Active Space	2u
Diameter of Peak Density	
Extra/Missing Metal	2u
Extra/Missing Poly	2u
Extra/Missing Active	2u
Maximum Defect Diameter	18u
Between Lot Alpha	100
Between Wafer Alpha	100
Wafers Per Lot	1
Radial Distribution	None
Minimum Line Spacing	0
Minimum Line Width	0

Table 6-1: DRAM Cell Process Conditions

sample. In Chapter 8 we describe realistic process conditions for use in simulating large circuits. The α values represent very little clustering between lots or wafers. Since there is only one wafer per lot, there is effectively only between-wafer clustering. The minimum line width and spacing are the Δw and Δs values discussed in Chapter 3 that specify the minimum line width and spacing that can be fabricated.

A simulator run of 1000 chip samples is shown in Figure 6-14. The output is a list of the unique chip fabrications. Each unique fabrication has a frequency count, and a list of fault groups. A fault group is a list of faults caused by a single defect. A small indentation signifies the continuation of a fault group.

For each fault, the fault type, defect type causing the fault, defect location, defect diameter, and fault information are provided. Some information has been omitted for clarity. In the case where several defects of the same type cause the same fault to occur, such as several oxide pinholes shorting the same nets together, the defect size and defect location are meaningless, since they only record the values for the first defect causing that fault. There is only one instance of this in the above list.

The result of 1000 simulated fabrications is 25 unique chip fault lists. The distribution of chip samples is shown in Table 6-2. The highly skewed nature of the distribution is typical of yield simulations. Excluding the chips with no circuit faults, 90% of the fault combinations are accounted for by only a few combinations. For more samples, this behavior holds true, while the tail lengthens dramatically. In a simulation of 10,000 samples of the three-transistor DRAM cell, 63% of the samples with faults (94% of the samples had no faults) were accounted for by 10 of the 118 unique fault lists. There were 69 fault lists that only occurred once, with up to 3 faults on a chip.

Despite the fact that an average of 2.5 defects were generated for each DRAM "chip", only 5.9% of all chips had a circuit fault. This result is typical of yield simulations. Only a small percentage of defects cause a circuit fault. For example, only 3.3% of the DRAM cell area contains gate oxide, so a gate oxide pinhole defect only has a 1 in 30 chance of causing a

```
SAM> vlasic -t omcellproc.dat omcell.pack
Reading wirelist omcell.pack
Writing faults to stdout
Parsing wafer file omcellwaf.cif
Initializing random number generators
left: -27.00 right: 3.00 bottom: -51.00 top: -3.00
NumXBins: 1 NumYBins 2
Allocating bins
Putting wirelist polygons into bins
Generating intermediate layers
Place and analyze defects
Results of fault testing:
Sample Count: 1000
Trial Count:
type POSM1: 234
type NEGM1: 237
type POSP: 283
type NEGP: 271
type POSD: 298
type NEGD: 335
type PIN1: 274
type PIN2: 276
type PING: 318
Total Trials: 2526
Total number of distinct chiplists: 25
Distinct Circuit Faults and Counts:

941 NIL

9 SHORT PIN1 X -10 Y -37 Diam 0 N3 N17

6 SHORT PIN1 X -25 Y -15 Diam 0 N2 N7

5 NEWVIA PING X -15 Y -16 Diam 0 NCHAN N7

5 SHORT PIN1 X -9 Y -17 Diam 0 N3 N7

4 SHORT PIN1 X -5 Y -21 Diam 0 N3 N10

4 NEWVIA PING X -13 Y -25 Diam 0 NCHAN N10

3 SHORT PIN1 X -20 Y -25 Diam 0 N2 N10

3 NEWVIA PING X -3 Y -37 Diam 0 NCHAN N17

3 SHORT PIN1 X -20 Y -37 Diam 0 N2 N17

2 SHORT POSP X 1 Y -27 Diam 16.09 N10 N17

2 SHORT POSP X -4 Y -12 Diam 7.43 N3 N7

1 SHORT POSM1 X -16 Y -14 Diam 17.19 N2 N3

1 SHORT POSP X -10 Y -43 Diam 9.04 N3 N17

1 SHORT PIN1 X -20 Y -37 Diam 0 N2 N17
  SHORT PIN1 X -25 Y -15 Diam 0 N2 N7

1 SHORT PIN1 X -10 Y -37 Diam 0 N3 N17
  SHORT PIN1 X -5 Y -21 Diam 0 N3 N10

1 NEWVIA PING X -15 Y -16 Diam 0 NCHAN N7
  SHORT PIN1 X -5 Y -21 Diam 0 N3 N10
  SHORT PIN1 X -20 Y -25 Diam 0 N2 N10

1 OPEN NEGP X -21 Y -41 Diam 12.25 N17/1 LEFT NP  N17/2 Tran D0 G

1 NEWVIA PING X -3 Y -37 Diam 0 NCHAN N17
  OPEN NEGM1 X -7 Y -43 Diam 7.58 N3/1 Tran D0 SD  N3/2 BOTTOM NM1 TOP NM1

1 OPEN NEGP X -7 Y -23 Diam 4.89 N10/1 Tran D0 SD  N10/2 Tran D1 G

1 OPEND NEGD X -14 Y -21 Diam 16.04 Tran D2
    OPEND NEGD X -14 Y -21 Diam 16.04 Tran D1

1 NEWGD POSP X -10 Y -19 Diam 8.43 CVTMULTI Tran D1 D2 SD: N2/0 N3/0 N9/0 G: N7/0 N10/0
    SHORT POSP X -10 Y -19 Diam 8.43 N7 N10

1 OPEND NEGD X -5 Y -35 Diam 4.91 Tran D0

1 SHORTD NEGP X -17 Y -15 Diam 8.10 Tran D2
    OPEN NEGP X -17 Y -15 Diam 8.10 N7/1 Tran D2 G  N7/2 LEFT NP

1 OPEN POSP X -24 Y -28 Diam 18.00 N2/1 Tran D1 SD LEFT ND  N2/2 LEFT NM1 BOTTOM NM1
    SHORT POSP X -24 Y -28 Diam 18.00 N2 N10 N17
```

Figure 6-14: VLASIC DRAM Example

```
94.1% No Faults
 4.2% One Oxide Pinhole Short
 0.6% One Extra Material Short
 0.2% Two Oxide Pinhole Shorts
 0.2% One Missing Material Open
 0.1% Three Oxide Pinhole Shorts
 0.1% Two Open Device
 0.1% One Open Device
 0.1% One Oxide Pinhole Short and One Missing Material Open
 0.1% One New Gate Device and One Extra Material Short
 0.1% One Shorted Device and One Missing Material Open
 0.1% One Extra Material Open and One Extra Material Short
```

Table 6-2: Chip Sample Distribution

gate to channel short. Recall from Chapter 4 that for a given set of design rules, extra and missing material defects must have a minimum diameter to cause a circuit fault. In the case of extra metal defects, with a diameter of peak frequency of 2 microns, and a minimum interesting diameter of 6 microns, only 1 in 18 defects has any chance of causing a short circuit. The smallest material defect that caused a fault in the example was 4.89 microns in diameter, which is larger than 91.7% of the material defects.

Another characteristic of the example that is common to all yield simulations is that chips with one simple fault are much more common than chips with several faults, or complicated fault groups. Single faults are more common than multiple faults because single defects are more common than multiple defects. Complicated fault groups are rare because large extra and missing material defects are required to cause them, and such large defects are rare. Of the four fault groups that occurred in the example, the smallest defect causing one was 8.1 microns in diameter. Only 1 in 33 defects is this large, so about 51 were generated during the simulation.

6.9.2. Static RAM Cell Example

A population of 1000 double-metal static RAM (SRAM) cells shown in Figure 6-15 was simulated using the process conditions described in Table 6-3. These conditions placed an average of 5.3 defects on each cell.

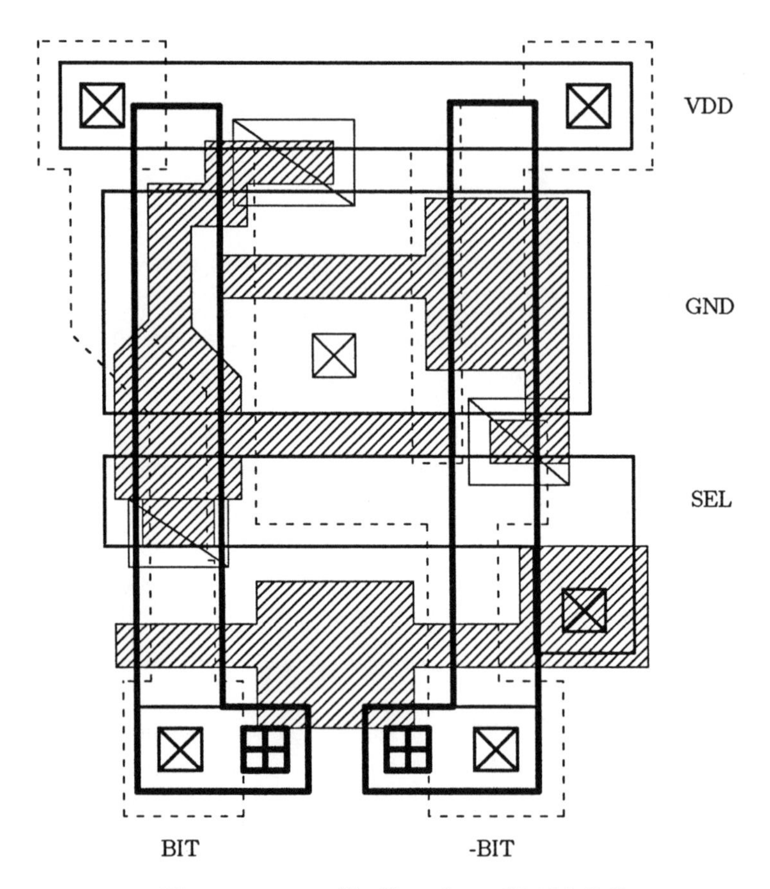

Figure 6-15: Six Transistor SRAM Cell

The results of this VLASIC simulation are shown in Figure 6-16. The listing contains only those unique fault lists with three or more occurrences (24 of 71 unique fault lists). As was the case with the DRAM example, the resulting circuit fault distribution is highly-skewed. There were only three circuit faults caused by extra or missing material defects. All other faults were caused by oxide pinholes or junction leakage. The reason for this is that the diameter of peak frequency for material defects was chosen to be 0.5 microns, while the minimum line widths in the SRAM cell are 3 microns. Thus only 1 in 432, or about 8 of the 3575 lithography defects are large enough to short or break a line. This relationship between diameter of peak

Defect Density	
Extra/Missing Metal1	20,000/cm2
Extra/Missing Metal2	20,000/cm2
Extra/Missing Poly	20,000/cm2
Extra/Missing Active	20,000/cm2
Second-Level Pinholes	20,000/cm2
First-Level Pinholes	20,000/cm2
Gate Oxide Pinholes	20,000/cm2
Design Rules	
Metal1 Width/Space	3u
Metal1 Contact Width	3u
Metal2 Width/Space	3u
Metal2 Contact Width	3u
Poly Width/Space	3u
Active Width/Space	3u
Poly/Active Space	1u
Diameter of Peak Density	
Extra/Missing Metal1	0.5u
Extra/Missing Metal2	0.5u
Extra/Missing Poly	0.5u
Extra/Missing Active	0.5u
Maximum Defect Diameter	18u
Between Lot Alpha	0.5
Between Wafer Alpha	4
Wafers Per Lot	50
Radial Distribution	None
Minimum Line Spacing	0
Minimum Line Width	0

Table 6-3: SRAM Cell Process Conditions

frequency and design rules is typical of many fabrication processes.

In a simulation of 10,000 samples of the SRAM cell, 74% of the samples with faults (72% of the samples had no faults) were accounted for by 27 of the 323 unique fault lists. There were 185 fault lists that only occurred once, with up to 7 faults on a chip.

6.10. VLASIC Performance

The DRAM example took 80 CPU seconds and 1.3 megabytes on a VAX-11/785 running Berkeley UNIX. This is 80 msec per chip sample, or 31.7 msec per placed defect. On a 10,000 sample simulation of the DRAM cell, the time falls to 72.3 msec per chip sample and 28.5 msec per placed defect, since the initialization overhead can be amortized over more chip samples. The DRAM cell has 30 input rectangles. An additional 21 are

```
gauss> vlasic -t sramtst.dat ramcell.pack
Reading wirelist ramcell.pack
Writing faults to stdout
Parsing wafer file /usr/dmw/yield/src/tmmonte/wafer.cif
Initializing random number generators
left: -2050.00 right: 2100.00 bottom: -3100.00 top: 2200.00
NumXBins: 2 NumYBins 2
Allocating bins
Putting wirelist polygons into bins
Generating intermediate layers
Place and analyze defects
Results of fault testing:
Sample Count: 1000
Trial Count:
type POSM1: 509
type NEGM1: 396
type POSP: 403
type NEGP: 482
type POSD: 427
type NEGD: 450
type POSM2: 511
type NEGM2: 397
type PIN1: 429
type PIN2: 436
type PING: 447
type PINJ: 460
Total Trials: 5347
Total number of distinct chiplists: 71
Distinct Circuit Faults and Counts:
705 NIL
35 SHORT PINJ X -2050 Y 2100 Diam 0 NSUB N6
23 SHORT PINJ X 750 Y -1300 Diam 0 NSUB N4
22 SHORT PIN1 X 150 Y 500 Diam 0 N3 N4
19 SHORT PINJ X -950 Y -600 Diam 0 NSUB N5
17 SHORT PIN2 X -950 Y 700 Diam 0 N1 N3
14 SHORT PIN2 X 1350 Y -400 Diam 0 NO N3
13 NEWVIA PING X -150 Y 700 Diam 0 NCHAN N4
12 NEWVIA PING X -450 Y -700 Diam 0 NCHAN N5
12 SHORT PINJ X -1050 Y -2600 Diam 0 NSUB N1
11 SHORT PINJ X 1150 Y -2600 Diam 0 NSUB NO
11 SHORT PIN1 X -1050 Y 1000 Diam 0 N3 N5
8 SHORT PINJ X 150 Y 400 Diam 0 NSUB N3
5 SHORT PIN2 X 750 Y -800 Diam 0 NO N2
5 NEWVIA PING X 650 Y -1900 Diam 0 NCHAN N2
5 SHORT PINJ X -2050 Y 2100 Diam 0 NSUB N6
  SHORT PINJ X -950 Y -600 Diam 0 NSUB N5
4 SHORT PIN2 X 1150 Y 1600 Diam 0 NO N6
4 SHORT PIN2 X -1250 Y 1500 Diam 0 N1 N6
  SHORT PIN2 X -950 Y 700 Diam 0 N1 N3
3 SHORT PIN1 X -50 Y 1500 Diam 0 N5 N6
3 SHORT PIN1 X -950 Y -1100 Diam 0 N2 N5
3 SHORT PINJ X -2050 Y 2100 Diam 0 NSUB N6
  SHORT PINJ X 750 Y -1300 Diam 0 NSUB N4
3 SHORT PIN2 X -950 Y -700 Diam 0 N1 N2
3 SHORT PIN1 X -350 Y -2500 Diam 0 N1 N2
3 SHORT PINJ X -2050 Y 2100 Diam 0 NSUB N6
  SHORT PIN2 X -950 Y 700 Diam 0 N1 N3
```

Figure 6-16: VLASIC SRAM Example

created by layer combination, or 0.7 per input rectangle.

The SRAM example took 1.3 megabytes and 151 CPU seconds or 151.2 msec per sample and 28.7 msec per defect. The time per defect is similar to that of the DRAM, despite the different technology, process, and layout. This indicates that the defect analysis time is relatively constant for typical layouts, at least for processes dominated by oxide pinhole defects. The SRAM cell has 180 input rectangles. This number is so large because the 45 degree edges are approximated by staircases in the Magic layout system. The generation of layer combinations creates another 174 rectangles, or about one per input polygon.

The breakdown in execution time for the 10,000 SRAM and DRAM sample simulations is given in Table 6-4.

	SRAM	DRAM
Polygon Operations	67.7%	81.6%
Random Number Gen	23.1	9.3
Main Control Loop	1.5	1.9
Fault Analysis	1.9	1.9
Fault Combination	0.8	1.0
Fault Summary	0.4	0.7
Memory Management	0.7	1.2
Other	3.9	2.4

Table 6-4: Execution Time Breakdown

The polygon operations and random number generation clearly dominate the execution time. The relative call frequency for the important polygon operations is shown in Table 6-5.

	SRAM	DRAM
ISECT	41.4%	68.8%
ENCLOSES	47.6	6.7
UNION	6.2	12.0
DIFFERENCE	4.1	10.4
INTERSECTION	0.7	2.1

Table 6-5: Polygon Operation Execution Breakdown

The frequency of usage varies strongly with the layout and the process conditions.

In the SRAM simulation, the top 8 procedures take 50% of the time, the top 21 take 75% of the time, and the top 41 take 90% of the time. The single most expensive operation is determining whether three points are colinear or not, as was shown in Figure 6-11, taking 15.6% of the total time. For the DRAM simulation, this calculation takes 20.7% of the time.

The time to analyze each defect that is placed on the DRAM cell is shown below in Table 6-6 along with the percentage of time spent in each processing phase.

	Time (msec)	Fault Analysis	Fault Combination	Polygon Operations
Extra Metal (S)	24.1	2.8%	4.5%	92.7%
Missing Metal (O)	101.1	1.8	1.3	96.9
Extra Poly (S/O/NGD)	257.1	1.1	0.6	98.3
Missing Poly (O/SD)	251.9	1.1	0.5	98.4
Extra Active (S/NAD)	63.4	2.3	2.2	95.5
Missing Active (O/OD)	157.8	1.6	0.9	97.5
Pinhole/Leakage (NV)	3.8	7.1	9.7	83.2

S = Short, O = Open SD ,= Shorted Device, OD = Open Device
NGD = New Gate Device, NAD = New Active Device, NV = New Via

Table 6-6: Defect Analysis Time

This data is for the process described in Table 6-1. The codes signify what types of faults the defect can cause.

The table makes clear that almost all of the processing time is taken up by polygon operations in the fault analysis routines. The table also shows that defects that cause open circuits are the most expensive to analyze, as are defects that can cause multiple faults. New via defects are by far the cheapest to analyze, using only an average of 4.54 ENCLOSES operations in the analysis.

The resulting faults are given in Table 6-7. In the SRAM process, most material defects are too small to cause a fault. Therefore the faults are primarily due to oxide pinholes. In the DRAM process, larger material defects are much more common, so the distribution of faults is not as skewed. For the SRAM simulation, only 4% of the extra and missing material defects generated were large enough to have any chance of causing a fault. For the DRAM, 23.8% of the extra and missing material defects could cause a fault. This is higher than for SRAM because in the SRAM process, material defects have a diameter of peak frequency of 0.5 microns, and a minimum line width of 3 microns, a 6:1 ratio. In the DRAM process, the peak is 2 microns and the minimum line width 4 microns, only a 2:1 ratio.

	SRAM	DRAM
Pinhole Short	85.8%	24.9%
Extra Material Short	1.7	21.3
Missing Material Open	0.6	29.6
Extra Material Open	0.2	5.3
Shorted Device	0.5	5.3
Open Device	0.0	5.3
New Gate Device	0.1	4.8
New Active Device	0.0	0.6
New Via	11.1	8.9

Table 6-7: Fault Distribution

The percentage of fault analyses that resulted in faults for each type of fault is given in Table 6-8.

	SRAM	DRAM
Short	1.2%	1.2%
Open	0.9	2.3
Shorted Device	6.9	2.5
Open Device	0.0	2.5
New Gate Device	0.2	0.6
New Active Device	0.0	0.1
New Via	4.5	0.7

Table 6-8: Fault Analyses Resulting In Faults

In general, a low percentage of defects result in faults, particularly the new device faults. These results show that the fault analysis procedures should be optimized to determine as quickly as possible whether a fault has occurred or not. If a fault does occur, then it can be analyzed.

6.10.1. Random Number Generator Performance

For the SRAM cell, 5.3 defects are placed per chip, with a random number generator time of 6.6 msec per defect. This is for a process with 50 wafers per lot, $\alpha = 0.5$ between lots, $\alpha = 4$ between wafers within a lot, and a 72% yield. For the DRAM, 2.5 defects are placed per chip, with a random number generator time of 2.6 msec per defect. The breakdown in execution time is given in Table 6-9.

	SRAM	DRAM
Poisson	18.5%	10.2%
Slow Negative Binomial	71.8	0.0
Fast Negative Binomial	4.3	75.5
Diameter	1.1	2.0
Uniform Integer	4.3	12.3

Table 6-9: Execution Breakdown

The fast negative binomial algorithm is used for small λ, and the slow algorithm is used for large λ. For the SRAM, 18.2% of the negative binomial deviates were generated using the fast algorithm. The fast algorithm was always used for the DRAM simulation due to the fact that $\alpha = 100$ and there was only one wafer per lot, so λ was low. The fast algorithm is 3.5 times faster on average than the slow algorithm used for large λ. The performance of the random number generators is shown in Table 6-10.

Negative Binomial	
alpha = 1, lambda = 4	0.4 msec/call
alpha = 1, lambda = 1000	1.5 msec/call
Poisson	
lambda = 10	0.7 msec/call
lambda = 1000	0.8 msec/call
Defect Size	0.1 msec/call
Random Integer (placement)	0.07 msec/call

Table 6-10: Random Number Generator Performance

We do not currently use the truncated defect size distribution described in Chapter 4. A truncated size distribution would reduce the random number generator time by reducing the number of calls to the size distribution, and by reducing the mean value of the negative binomial and Poisson distributions. The diameter calculation already takes a small percentage of the time. The Poisson distribution is used to determine the number of defects in a wafer zone. If there are on average 0.5 defects of each type per chip, then a 50-die zone will have an average of 25 defects. A truncated size distribution could reduce this value by a factor of 72. This reduces the time for a Poisson deviate by a factor of seven. This reduces the random number generator time by 9-16%. For 0.5 defects per chip and 50 wafers per lot, there are on average 50 defects per wafer and 2500 defects per lot. Reducing them by a factor of 72 allows use of the fast algorithm for the between-wafer negative binomial but still requires the slow algorithm for the between-lot negative binomial. This reduces the negative binomial time by 26%. The total time reduction in the random number generators would be 42%. This would reduce the total yield simulation time by 10%.

6.11. Simulation Accuracy

Apart from inaccuracies in the defect models and statistics, a VLASIC simulation has inaccuracies inherent to the Monte Carlo method. A VLASIC simulation is equivalent to sampling from a multinomial distribution with n chip samples, t fault group combination categories, category probabilities $\bar{p}=(p_1 \ldots p_t)$, and category counts $\bar{x}=(x_1 \ldots x_t)$. The probability function is

$$f_m(\bar{x} \mid n, \bar{p}) = n! \prod_{i=1}^{t} \frac{p_i^{x_i}}{x_i!}, \tag{6-1}$$

$$\sum_{i=1}^{t} p_i = 1, \qquad 0 \leq p_i \leq 1. \tag{6-2}$$

The category counts resulting from a simulation are $\bar{y}=(y_1 \ldots y_t)$. The maximum likelihood estimates (MLE) for the p_i are

$$\pi_i = \frac{y_i}{n}. \tag{6-3}$$

The exact number of fault group combinations t is unknown but extremely large. VLASIC only outputs a category if it has at least one count.

Since each category is mutually exclusive, we can combine several categories together, summing their x_i and p_i and appropriately adjusting t, to form a new multinomial. The redundancy analysis system described in Chapter 7 is an example of recategorization.

There are several possible measures of simulation accuracy, depending on the application. For any one category, we can determine confidence intervals for the category count, or the category probability, or conversely, determine the sample size required to achieve a given confidence interval. The random number generators can also be a source of error.

This analysis can be extended to determine the simultaneous confidence intervals and halting criteria for several category counts or probabilities. This can be done by considering $\bar{x}$ and $\bar{p}$ to be points in a t-dimensional count or probability space, and determining the confidence volume [Walker 86a].

6.11.1. Confidence Intervals for Category Counts

The confidence interval for the category count x_i for an individual category i may be determined by first combining all other categories together. The VLASIC simulation can then be considered a sequence of n Bernoulli trials with probability p_i. The category count x_i is then given by the binomial distribution with probability function

$$f_b(x_i \mid n, p_i) = \binom{n}{x_i} p_i^{x_i} (1-p_i)^{n-x_i}, \qquad 0 \leq x_i \leq n. \tag{6-4}$$

The $100(1-\mu)$ percent confidence interval for x_i is $c_{i,1} \leq x_i \leq c_{i,2}$ where $c_{i,1}$ and $c_{i,2}$ are chosen so such that

$$\sum_{x_i = c_{i,1}}^{c_{i,2}} f_b(x_i \mid n, \pi_i) = 1 - \mu . \tag{6-5}$$

The interval endpoints $c_{i,1}$ and $c_{i,2}$ are chosen to minimize $(c_{i,2} - c_{i,1})$.

6.11.1.1. Approximations to the Binomial Distribution

For large n, Equation (6-5) is difficult to compute. If y_i is small and n large, then the binomial can be approximated by the Poisson distribution with probability function

$$f_p(x_i \mid n, p_i) = \frac{e^{-np_i} (np_i)^{x_i}}{x_i!}, \qquad 0 \leq x_i \leq n. \tag{6-6}$$

The confidence interval $c_{i,1} \leq x_i \leq c_{i,2}$ is determined by solving

$$\sum_{x_i = c_{i,1}}^{c_{i,2}} f_p(x_i \mid n, \pi_i) = 1 - \mu . \tag{6-7}$$

This approximation is still difficult to compute.

If y_i is large enough, then the binomial distribution can be approximated by the normal distribution. The $100(1-\mu)$ percent confidence interval for the category count is determined by the two-sided t-test. The interval is

$$y_i - \frac{\sqrt{y_i(n-y_i)}}{\sqrt{n-1}} t_{\mu/2;n-1} \leq x_i \leq y_i + \frac{\sqrt{y_i(n-y_i)}}{\sqrt{n-1}} t_{\mu/2;n-1} \tag{6-8}$$

where $t = 1.96$ for $\mu = 0.05$ and $n > 120$ [Lukacs 72, DeGroot 75]. If the lower bound is negative, it is truncated to zero.

The confidence interval can also be approximated using the Pearson X^2 statistic

$$X^2(\bar{y}, \bar{x}) = \sum_{i=1}^{t} \frac{(y_i - x_i)^2}{x_i} . \tag{6-9}$$

X^2 has a chi-square distribution with $t-1$ degrees of freedom $\chi^2_{\mu;t-1}$. The $100(1-\mu)$ percent confidence interval is determined by solving

$$\frac{(y_i - x_i)^2}{x_i} + \frac{[(n-y_i) - (n-x_i)]^2}{n - x_i} = \chi^2_{\mu;1} \tag{6-10}$$

for x_i. A negative lower bound is truncated to zero. For $\mu = 0.05$, and one degree of freedom, $\chi^2 = 3.841$.

For $n = 10{,}000$, the Poisson, normal, and Pearson approximations are compared with the 95% binomial confidence interval in Table 6-11. Note that since the binomial and Poisson equations are integer-valued, the confidence interval endpoints are integers. The results indicate that for $y_i \leq 5$, the Poisson approximation is exact, while the normal approximation is slightly inaccurate. For $y_i > 5$, the normal approximation is the most

Category Count	Binomial	Poisson	Normal	Pearson
	Confidence Intervals			
0	- to --	- to --	-.-- to --.--	0.00 to 3.84
1	0 to 3	0 to 3	0.00 to 2.96	0.18 to 5.66
2	0 to 4	0 to 4	0.00 to 4.77	0.54 to 7.29
3	0 to 6	0 to 6	0.00 to 6.39	1.02 to 8.82
4	0 to 7	0 to 7	0.08 to 7.92	1.56 to 10.28
5	1 to 9	1 to 9	0.62 to 9.38	2.14 to 11.70
10	4 to 16	3 to 16	3.80 to 16.20	5.43 to 18.40

Table 6-11: 95% Confidence Intervals for 10,000 Samples

accurate. The X^2 statistic approaches the normal approximation as y_i increases. The table values do not change significantly for $n = 1000$.

For the example in Figure 6-14, $y_0 = 941$. The 95% confidence interval for the category count is $926.4 \leq x_0 \leq 955.6$. The next most common category has a count $y_1 = 9$. The 95% confidence interval is $3.1 \leq x_1 \leq 14.9$. The intervals for the other example categories can be calculated from Table 6-11.

6.11.2. Confidence Intervals for Category Probabilities

The confidence interval for the category probability p_i for category i can be calculated by treating the simulation as a sequence of Bernoulli trials as was done in Section 6.11.1. If we assume that the prior distribution of p_i is uniform, i.e. we make no assumptions about p_i, then the posterior distribution is a beta distribution with parameters $\alpha_i = 1 + y_i$ and $\beta_i = n + y_i - 1$ [DeGroot 75]. The probability density function of the beta distribution is

$$\xi(p_i \mid y_i) = \frac{(n+1)!}{y_i!(n-y_i)!} p_i^{y_i} (1-p_i)^{n-y_i}, \qquad 0 \leq p_i \leq 1. \tag{6-11}$$

The $100(1-\mu)$ percent confidence interval for p_i is $\pi_{i,1} \leq p_i \leq \pi_{i,2}$ where $\pi_{i,1}$ and $\pi_{i,2}$ satisfy the relationship

$$\int_{\pi_{i,1}}^{\pi_{i,2}} \xi(p_i \mid y_i) dp_i = 1 - \mu. \tag{6-12}$$

Integrating, we get

$$-(1-p_i)^{n-y_i+1} \sum_{j=0}^{y_i} \frac{p_i^j (n-y_i+j)!}{j!(n-y_i)!} \Big|_{\pi_{i,1}}^{\pi_{i,2}} = 1 - \mu\,. \tag{6-13}$$

Ideally $\pi_{i,1}$ and $\pi_{i,2}$ are chosen to minimize $(\pi_{i,2} - \pi_{i,1})$. Since the beta distribution is not symmetric, the ideal values will not be centered about the mean. This is obviously true when y_i is small, since $\pi_{i,1} = 0$ in this case. To simplify our analysis, we always assume that $\pi_{i,1}$ and $\pi_{i,2}$ are symmetrical about the distribution mean, except where $\pi_{i,1} = 0$. Sensitivity analysis indicates that the symmetric confidence intervals are only slightly longer than the minimum intervals for typical yield simulations.

6.11.2.1. Approximations to the Beta Distribution

Equation (6-13) is difficult to solve. We can approximate the beta distribution with a normal distribution $\Phi(\lambda_i, \sigma_i^2)$ and use the *t*-test to obtain the confidence interval

$$\lambda_i - \sigma_i t_{\mu/2;n-1} \leq p_i \leq \lambda_i + \sigma_i t_{\mu/2;n-1} . \tag{6-14}$$

We use two different ways of determining λ_i and σ_i. Our first approximation uses the expected mean and standard deviation for the beta distribution

$$\lambda_i = \frac{y_i + 1}{n + 2} , \tag{6-15}$$

$$\sigma_i^2 = \frac{(y_i + 1)(n - y_i + 1)}{(n + 2)^2(n + 3)} . \tag{6-16}$$

We call this the *normal-beta* approximation. The second approximation uses the expected values from the binomial sample distribution

$$\lambda_i = \pi_i ,$$

$$\sigma_i^2 = \frac{\pi_i (1 - \pi_i)}{n - 1} .$$

We call this the *normal-binomial* approximation. Negative interval endpoints resulting from the approximations are truncated to zero.

The approximations are compared to the beta distribution for $n = 10{,}000$ and $\mu = 0.05$ in Table 6-12. Note that for small y_i, the normal-binomial interval for y_i is the same as the normal-beta interval for $y_i - 1$. The

Category Count	Confidence Intervals Beta	Normal-Beta	Normal-Binomial
0	.00000 to .00030	.00000 to .00030	.----- to .-----
1	.00000 to .00048	.00000 to .00048	.00000 to .00030
2	.00000 to .00063	.00000 to .00064	.00000 to .00048
3	.00002 to .00078	.00001 to .00079	.00000 to .00064
4	.00008 to .00092	.00006 to .00094	.00001 to .00079
5	.00014 to .00106	.00012 to .00108	.00006 to .00094
10	.00047 to .00174	.00045 to .00175	.00038 to .00162

Table 6-12: 95% Confidence Intervals for 10,000 Samples

Category Count	Relative Error Normal-Beta	Normal-Binomial
0	0.0%	--.-%
1	+0.1%	-37.7%
2	+1.6%	-24.2%
3	+3.8%	-15.3%
4	+4.3%	- 6.7%
5	+4.1%	- 4.6%
10	+2.3%	- 2.4%

Table 6-13: Approximation Error for 10,000 Samples

relative errors of the confidence interval lengths are shown in Table 6-13.

For $n = 1000$, the beta intervals and the approximations are shown in Table 6-14.

Category Count	Confidence Intervals Beta	Normal-Beta	Normal-Binomial
0	.00000 to .00299	.00000 to .00295	.----- to .-----
1	.00000 to .00473	.00000 to .00476	.00000 to .00296
2	.00000 to .00628	.00000 to .00638	.00000 to .00477
3	.00025 to .00773	.00009 to .00789	.00000 to .00639
4	.00081 to .00917	.00063 to .00935	.00009 to .00791
5	.00140 to .01058	.00121 to .01076	.00063 to .00937
10	.00468 to .01728	.00453 to .01743	.00383 to .01617

Table 6-14: 95% Confidence Intervals for 1000 Samples

Note that the interval values are approximately ten times as large for the 1000 sample case as the 10,000 sample case. The error for the approximations is shown in Table 6-15. The table indicates that the approximations maintain their accuracy for small sample sizes.

As the category count approaches $n/2$, the error in confidence interval length goes to zero.

Category Count	Relative Error Normal-Beta	Normal-Binomial
0	-1.2%	--.-%
1	+1.0%	-37.4%
2	+1.5%	-24.0%
3	+4.3%	-14.6%
4	+4.3%	- 6.4%
5	+4.0%	- 4.7%
10	+2.4%	- 2.1%

Table 6-15: Approximation Error for 1000 Samples

For the example in Figure 6-14, $\pi_0 = 0.941$. The 95% confidence interval for the probability is $0.925 \leq p_0 \leq 0.955$. The next most common category probability is $\pi_1 = 0.009$. The 95% confidence interval is $0.0038 \leq p_1 \leq 0.016$. The other category confidence intervals can be calculated from Table 6-14.

For most applications, the normal-beta approximation is adequate for approximating the 95% confidence interval. In some applications, the approximation may not be sufficiently accurate. For example, when the yield of a cell is used to determine the yield of a chip, a small error in the cell yield confidence interval causes a large error in chip yield confidence interval. If a cell is replicated 1000 times, a 0.01% error in the cell yield causes a 10% error in the array yield. This is discussed further in Chapter 9. For higher confidence levels, these approximations begin to break down, because the beta distribution has a longer tail than the normal distribution.

6.11.3. Halting Criteria

The VLASIC simulation can halt when the desired accuracy is obtained. For individual categories, accuracy can be defined as absolute confidence interval lengths, relative confidence interval lengths (length divided by the mean), the variance, or as the relative variance (variance divided by the square of the mean). The relationship between sample size and accuracy is used to determine when to halt simulation.

6.11.3.1. Halting Criteria for Category Counts

The variance of an individual category count is

$$VC_i = \frac{y_i(n - y_i)}{n - 1}. \tag{6-17}$$

The worst-case variance occurs when $y_i = n/2$, and increases with n. Hence VC_i is not a suitable measure of accuracy. The relative variance is the variance over the square of the mean (the square of the coefficient of variation)

$$VCR_i = \frac{n - y_i}{y_i(n - 1)}. \tag{6-18}$$

The relative variance is infinite for $y_i = 0$. It is not possible to a priori determine the sample size required to achieve a given VCR_i. The simulation can halt when y_i and n satisfy Equation (6-18). If the true category probability is p_i, then the expected number of samples n is given by

$$n = \frac{1 - p_i}{p_i VCR_i} + 1. \tag{6-19}$$

For $p_i = 0.0001$ and $VCR_i \leq 0.1$, $n \geq 99991$.

The length of the confidence interval for the category count is

$$LC_i = 2\sqrt{VC_i}\, t_{\mu/2;n-1}\,. \tag{6-20}$$

For $n > 120$, $t_{\mu/2;n-1}$ is constant, so LC_i is proportional to $\sqrt{VC_i}$. Since VC_i is unsuitable as an accuracy measure, LC_i is also unsuitable. The relative confidence interval (length over the mean) is

$$LCR_i = 2\sqrt{VCR_i}\, t_{\mu/2;n-1}\,. \tag{6-21}$$

For category probability p_i, the expected number of samples is

$$n = \frac{4(1 - p_i)t_{\mu/2;n-1}}{p_i LCR_i}, \tag{6-22}$$

assuming $n > 120$ so that $t_{\mu/2;n-1}$ is a constant. For $p_i = 0.0001$, $\mu = 0.05$, and $LCR_i \leq 1$, $n \geq 78393$.

6.11.3.2. Halting Criteria for Category Probabilities

The variance of an individual category probability, assuming a uniform prior distribution, is

$$VP_i = \frac{(y_i + 1)(n - y_i + 1)}{(n+2)^2(n+3)}. \qquad (6\text{-}23)$$

The worst-case variance occurs when $y_i = n/2$. For a worst-case variance $VP_i \leq 0.0001$, $n \geq 2497$. The relative variance is

$$VPR_i = \frac{(n - y_i + 1)}{(y_i + 1)(n+3)}. \qquad (6\text{-}24)$$

The worst-case relative variance occurs when $y_i = 0$. As was the case for category counts, it is not possible to a priori determine n for a desired VPR_i. The simulation can halt when y_i and n satisfy Equation (6-24). For category probability p_i, the expected number of samples is determined by solving

$$VPR_i = \frac{(n - np_i + 1)}{(np_i + 1)(n+3)} \qquad (6\text{-}25)$$

for n. For $p_i = 0.0001$ and $VPR_i \leq 0.1$, $n \geq 89988$.

If we use the normal-beta approximation for calculating the confidence interval, then the confidence interval length is

$$LP_i = 2\sqrt{VP_i}\, t_{\mu/2;n-1}. \qquad (6\text{-}26)$$

For the worst-case interval $LP_i \leq 0.01$ and $\mu = 0.05$, $n \geq 38414$. The relative interval length is

$$LPR_i = 2\sqrt{VPR_i}\, t_{\mu/2;n-1}. \qquad (6\text{-}27)$$

For category probability p_i, the expected number of samples is determined by solving Equations (6-25) and (6-27) for n. For $p_i = 0.0001$, $\mu = 0.05$, and $LPR_i \leq 1$, $n \geq 146109$.

6.11.4. Random Number Generator Accuracy

One possible source of simulation error is in the random number generators. The simulator uses a uniform distribution, a Poisson distribution, and a negative binomial distribution. The negative binomial distribution is implemented with a gamma distribution and Poisson distribution. The uniform distribution is highly accurate. The negative binomial distribution accuracy is determined by the Poisson and gamma distribution accuracy.

For one million samples, the error in mean and variance of the Poisson distribution generator is shown in Table 6-16. The expected mean and variance are λ.

Lambda	Sample Mean	95% Interval
10	9.9976	9.9914 to 10.0004
1000	999.9851	999.9231 to 1000.0471
10000	9999.9176	9999.7214 to 10000.1138
100000	99999.7304	99999.1098 to 100000.3510

Lambda	Sample Variance	95% Interval
10	10.0313	10.0040 to 10.0586 **
1000	1002.2193	999.4898 to 1004.9620
10000	10025.1531	9997.8504 to 10052.5880
100000	100245.5347	99972.5345 to 100519.8781

** within 99% confidence interval

Table 6-16: Poisson Distribution Error

The distribution uses different algorithms for the ranges $0 < \lambda \leq 10$ and $\lambda > 10$. For the between-lot and between-wafer distributions, the Poisson mean is in the thousands, where it is accurate. For determining the number of defects to place on the chip, the mean is less than 10, where the variance is slightly high.

For one million samples, the error in the mean and variance of the gamma distribution generator is shown in Table 6-17. The expected mean and variance are λ. The generator uses different algorithms for the ranges $0 < \lambda \leq 1.0$, $1 < \lambda \leq 2.53$, and $\lambda > 2.53$. For typical yield simulations, the mean is in the range 0.1 to 100, where it is fairly accurate.

Lambda	Sample Mean	95% Interval			
0.5	0.5007	0.4993	to	0.5021	
2.5	2.5048	2.5017	to	2.5079	*
10.0	10.0102	10.0040	to	10.0164	*
1000.0	1000.2356	1000.1736	to	1000.2976	*
10000.0	10000.7801	10000.5830	to	10000.9772	*

Lambda	Sample Variance	95% Interval			
0.5	0.5009	0.4995	to	0.5023	
2.5	2.4999	2.4931	to	2.5067	
10.0	10.0338	10.0064	to	10.0612	**
1000.0	1001.7742	999.0459	to	1004.5157	
10000.0	10107.6023	10080.0750	to	10135.2628	*

* outside 95% confidence interval
** within 99% confidence interval

Table 6-17: Gamma Distribution Error

6.12. Sensitivity Analysis

The simulation results differ with different defect statistical models and their parameters. In some cases, different models result in the same simulation results, so the most convenient model is selected. An example of this is the use of the truncated defect size distribution in Chapter 4. The shape of the distribution is irrelevant for defects too small to cause a circuit fault, so it is simply chopped to zero. In other cases, the distribution can make a significant difference to the simulation results.

6.12.1. Statistical Model Sensitivity

The simulation results may vary with the defect size or spatial distribution. For example, the simulation results are sensitive to the distribution of defect sizes for defects large enough to cause faults. The distribution x_0/x^2 has S/x_0 times as many defects larger than diameter S as the distribution x_0^2/x^3. If $x_0 = 0.5$ microns and $S = 10$ microns, then the difference is a factor of 20. If these distributions represented the size distribution of extra material defects, and S was the spacing between long parallel lines, then the choice of distribution can affect the frequency of short circuits by a factor of 20.

The simulation results are relatively insensitive to the radial distribution of defects on the wafer. A least-square fit of radial distributions to the process monitor data in Chapter 8 shows that several distributions, such as piecewise-linear, two-zone, and exponential provide almost equally good fits, and so will result in almost the same number of defects being placed on a die.

Variations in the between-lot and between-wafer defect distributions result in substantial variations in the simulation results. If a single distribution over all lots is used to fit data with high between-wafer defect clustering, then the distribution is less skewed than the data. The distribution will predict more wafers with no defects and more wafers with many defects than there really are. As an example, assume a set of ten lots with five wafers each. Six lots have $\lambda_w = 100$ and $\alpha_w = 1$. Four lots have $\lambda_w = 10$ and $\alpha_w = 1$. The between-lot distribution derived from this data has $\lambda_l = 230$ and $\alpha_l = 1.09$. A single distribution fit to all 50 wafers has $\lambda = 46$ and $\alpha = 0.53$. The distribution of defects is shown in Table 6-18.

Defects/ Wafer	Percent of Wafers with Defect Count Two-Level Distribution	One-Level Distribution
> 0	93.15%	74.22%
> 20	47.75%	38.41%
> 40	31.24%	28.06%
> 100	12.65%	14.21%

Table 6-18: High Between-Wafer Clustering Distributions

If there is little between-wafer clustering, then a single distribution will fit the data. If we assume that the between-wafer distribution is Poisson ($\alpha_w = \infty$), then a one-level distribution will match a two-level distribution. A negative binomial distribution is a Poisson distribution with a gamma-distributed mean, so a negative binomial distribution followed by a Poisson distribution is a negative binomial distribution. A slight error is introduced because the between-lot distribution has $\lambda_l = n\lambda_w$ where n is the number of wafers per lot. The resulting defect count is divided by n to get the expected number of defects per wafer used by the between-wafer Poisson distribution. Similarly, if there is little between-lot clustering, then a single wafer-level negative binomial distribution will fit the data.

The above results imply that when it is difficult to select from several candidate distributions to fit the data, then they will all result in similar simulation results. When the simulation is particularly sensitive to a change in distribution, then one distribution will clearly fit the data best [Stapper 84a].

6.12.2. Parameter Sensitivity

When α is small, small changes in its value result in substantial changes in the number of chips with no defects. A 10% change in α can easily cause a 30% change in yield. When α is large, then the distribution is nearly Poisson, and the yield is insensitive to large changes in α.

Changes in λ result in small changes in the number of wafers with $\ll \lambda$ defects, a proportional change in the number of wafers with $\approx \lambda$ defects, and a large change in the number of wafers with $\gg \lambda$ defects. For the single-level negative binomial distribution above, a 10% change in λ results in a 0.5% change in the number of wafers with no defects, a 7.4% change in wafers with > 46 defects, and a 27.4% change in wafers with > 200 defects.

As was the case for the distribution shape, those parameter changes that result in large changes in simulation results also lead to sharp minima when fitting the distributions to the process monitor data.

6.12.3. Fault Detection Sensitivity

Some types of defects rarely occur, but are expensive to detect. For example, the new active device procedure adds about 40 msec to the processing time for an extra active defect, but in a simulation of an array of DRAM cells, only 40 of the 6536 resulting faults groups contained new active devices. This represents 5.7% of the total simulation time for 0.6% of the faults. Similarly only 123 of the fault groups contained new gate devices, but their detection adds 127 msec to the processing time for an extra poly defect. This is 18.0% of the total time for 1.9% of the faults.

If we did not try to detect the new active and new gate devices, we could reduce the simulation time by 23.7% while introducing only a 2.5% error into the analysis. If there are no other faults in the fault groups, then the number of fault groups will decrease by 2.5%. Since many of the chips with new devices contain other fault groups, the number of chips with faults will not change significantly. In fact 20 of the 40 new active devices and 115 of the 123 new gate devices occur in fault groups that also contain shorts or opens, so the number of fault groups will decrease by only 0.4%.

All of the other types of circuit faults are common enough that ignoring them would introduce substantial errors into the analysis.

Chapter 7
Redundancy Analysis System

This chapter describes the application of VLASIC for analyzing the effects of adding redundancy to a design. Redundancy analysis is accomplished by means of a post-processor that uses the chip fault lists generated by VLASIC and a description of chip redundancy to predict yield in the presence of redundancy. We first discuss previous work in predicting the yield of redundant chips, then describe our redundancy analysis system, and finally provide examples of its use.

By using VLASIC and a redundancy analysis post-processor (RAPP), we can build an automated redundancy analysis system that makes accurate yield predictions. VLASIC automatically produces a distribution of chip fault lists matching those observed in the fabrication process. The redundancy analysis post-processor then classifies these faults into the categories of interest, such as single-bit, double-row, or column failures, and determines whether the available redundancy can cover the fault. By splitting the redundancy analysis system into two parts and taking advantage of the modularity of VLASIC, it is possible to extend the system to other defect types, defect statistics, and redundancy techniques.

7.1. Previous Work on Redundancy Analysis

Redundancy has been shown to be an effective technique for increasing integrated circuit yield, particularly in DRAMs [Fitzgerald 80, Mano 80, Abbott 81, Smith 81a, Smith 81b, Benevit 82, Fitzgerald 82, Ishihara 82, Chwang 83], SRAMs [Hardee 81, Ebel 82, Minato 82, Smith 82], and EPROMs [Yoshida 83]. In a few cases, error-correction has been used instead of redundancy, to improve yield [Yamada 84, Davis 85]. Several methods have been proposed for predicting the improvement in yield that would result through the use of redundancy. We begin by describing some of the methods and their shortcomings.

7.1.1. Yield Models with Redundancy

Yield models that have been developed for predicting the yield of chips without redundancy (see Chapter 2) can be extended to chips that include redundancy. Rather than just predicting the number of chips on which a circuit fault occurred, one could predict the number and type of faults. The number of faults can be provided by the defect distribution under the assumption that the number of faults is proportional to the number of defects. Since redundancy can only repair certain kinds of failures, it is necessary to categorize the faults that occur. For the case of redundant memories, fault categories include bit, row, and column failures. The relative probabilities of these faults can be determined empirically, or by estimating the critical area for each type of fault.

Boltzmann or Poisson statistics have been used to predict the yield of memories with error correction or redundancy for SRAMs, DRAMs, and wafer-scale integration, and to select the appropriate amount of redundancy [Borisov 78, Borisov 79, Sud 81, Fujii 83, Mano 82, Sakurai 84, Uchida 82, Egawa 80, Kitano 80, Ueoka 84, Raffel 85]. These statistics were also used to predict the yield and redundancy requirements for transistor arrays, logic blocks, memories, and systolic arrays, some with discretionary wiring [Wallmark 60, Hofstein 63, Tammaru 67, Chen 69, Moore 84a, Moore 84b]. As was discussed in Chapter 2, the defects that occur in the fabrication line do not obey Poisson statistics. Defect clustering results in higher yields than predicted by the Poisson distribution, so for chips without redundancy, Poisson statistics are pessimistic. Poisson statistics are optimistic for chips with redundancy. The Poisson distribution predicts fewer chips with many faults than actually occur [Stapper 80]. This leads to the conclusion that a given amount of redundancy can correct a larger fraction of failures than is actually the case. It has been reported that the misuse of Poisson statistics has resulted in commercial disaster [Peltzer 83, Stapper 85, Stapper 86].

Recognizing that Poisson statistics are inadequate, researchers considered Bose-Einstein and mixed Poisson statistics. Both types of statistics have been used to determine the amount of necessary redundancy and to predict RAM yield [Cenker 79, Ochii 82]. Bose-Einstein statistics have also been used to analyze redundant wafer-scale logic arrays and systolic arrays [Harden 86, Mangir 82, Mangir 84, McCanny 83]. These techniques allowed the designers to make rough evaluations of design alternatives, but not to make accurate yield predictions.

As discussed in Chapter 4, negative binomial statistics have been found to best model the defect clustering that occurs during fabrication. Negative binomial statistics have been used for selecting the number of spare rows and columns for redundant RAMs [Khurana 82, Stapper 82c, Haraszti 82], and PROMs [Gongwer 83], and for analyzing a wafer-scale processor array [Koren 84]. These statistics were successful in predicting yields and redundancy requirements for memory arrays manufactured in a well-characterized fab line.

Ketchen used negative binomial statistics to model both intrawafer and interwafer defect density variations when analyzing wafer-scale integration [Ketchen 85]. Interwafer variations had a much stronger influence on yield than intrawafer variations. Local correlations between defects on a wafer were also considered, and a sharp drop in yield was found when the spacing between defects was similar to the spacing between a functional and spare unit. Stapper reemphasized the importance of interwafer defect clustering when analyzing wafer-scale integration [Stapper 85].

These analytic yield models had several shortcomings. First, the analysis usually estimated the critical area based on the average defect size or empirical data. This inaccuracy will be demonstrated in an example below. Second, when accurate critical area estimates were made, they were usually done by hand, which can be quite time consuming. Third, all faults were treated as fatal for those cells in which they occurred. This is not always the case, particularly in fault-tolerant cells. For example, a signal may be carried on several lines, so that a break in any one line does not cause an open circuit. Some of these problems, such as poor critical area estimation, were solved with Monte Carlo analysis, described in the next section.

7.1.2. Yield Simulation with Redundancy

An alternative to analytic techniques for yield prediction are methods that employ simulation. Stapper et al. estimated the probability of a defect causing a failure (bit, row, column, etc) on a DRAM using Monte Carlo simulation, where defects were placed on a checkplot (by hand) and visually examined [Stapper 80, Stapper 82b]. The defects were drawn from the $1/x^3$ size distribution described in Chapter 4. The failure probabilities were combined with the cell areas and negative binomial statistics to determine the probability of different failure combinations. By using knowledge of what combinations could be repaired, the yield in the presence of redundancy was determined. Schuster performed a simplified version of this analysis using a $1/x^2$ size distribution and Poisson statistics [Schuster 78]. These techniques could make very accurate yield predictions, but were quite time consuming, and thus used only for high-volume memory products.

York implemented a hybrid system to estimate the yield and reliability of triple modular redundancy wafer-scale integration [York 85]. Individual element failure probabilities were calculated using both a simple yield model with negative binomial statistics, and by a method using estimated critical areas and a Monte Carlo simulation. A similar analysis was done for inter-element wiring. A Monte Carlo simulation was done with elements and wiring failing with these probabilities. The failures were then automatically analyzed to determine whether the static redundancy would cover the failure. This technique was successfully used to evaluate the economic feasibility of wafer-scale integration.

Barton used a pure simulation technique to estimate the yield of a fault-tolerant RAM [Barton 80a]. Defects were drawn from a $1/x$ size distribution, uniformly placed on the layout, and visually examined. Neither the size or spatial distribution reflect reality. The analysis was also quite time consuming.

Kung and Lam used a simulation technique to predict the yield of fault-tolerant wafer-scale systolic arrays [Kung 84]. The simulator assumed that systolic processors had some probability of failing with a uniform spatial distribution. Registers and wiring interconnecting systolic processors were assumed not to fail. It is questionable whether such an assumption would hold in reality.

The simulation techniques that made use of the correct defect statistics were able to make accurate yield predictions. The drawbacks were that the techniques were quite time-consuming and not readily extended to other applications. We will come back to these techniques in Chapter 9 and how they might be made more flexible.

7.2. Redundancy Analysis Post-Processor

The structure of the redundancy analysis post-processor is shown in Figure 7-1.

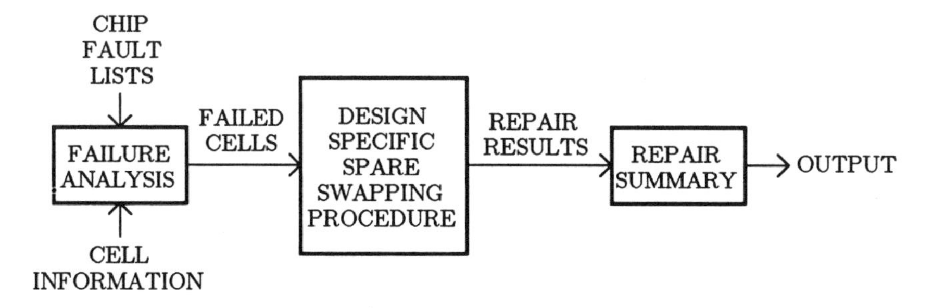

Figure 7-1: Redundancy Analysis System Structure

The post-processor is divided into three main modules: failure analysis, spare swapping, and repair summary. The failure analysis module takes the unique chip fault lists generated by VLASIC and uses cell descriptions and a set of heuristics to determine what cells have failed. These failed cells are then passed to the spare swapping procedure. The spare swapping procedure uses a spare swapping algorithm, and the knowledge of what cells are spare, what cells have spares, and what cells do not have spares to determine whether a set of cell failures can be corrected or not. The repair results are stored in a repair summary module for output when the simulation is complete.

Ideally the redundancy analysis system should be completely general purpose. The failure analysis module should be able to take an arbitrary design description and use general-purpose failure analysis heuristics to determine what cells have failed due to a circuit fault. The design-specific spare swapping and summary modules should be constructed to accept a list of failed cells. The system that we have implemented is not completely general purpose. The failure analysis module is designed for evaluating failures in memories. The cell information file format is specialized for memories. Failures are described in terms of bits, rows, and columns. The spare swapping and summary modules have been designed to take advantage of this specialization.

We first describe the failure analysis module and the heuristics that it uses. We then describe the spare swapping procedure for a memory. Note that we use the term "failure analysis" to mean determining what cells in a chip no longer function properly, while we use the term "fault analysis" to mean determining what changes to the DC circuit topology have occurred due to a defect.

7.2.1. Failure Analysis

The failure analysis module takes the set of unique chip fault lists generated by VLASIC and cell descriptions as input. The unique fault lists are used rather than the direct stream of fault lists available inside of VLASIC because this saves work. A fault list that appears several times need only be analyzed once. In the example below, this saves 8.4% of the work. VLASIC uses fully-instantiated geometry, and so has no concept of cells.

The unique fault lists describe faults in terms of global identifiers for transistors and nets. The cell description provides a mapping between these identifiers and individual cells. This mapping is necessary since spares and spare swapping are defined in terms of cells, rather than individual transistors and nets. The description also includes the interconnection between cells. This interconnection information is necessary since a fault in a net connecting several cells together can cause the failure of all those cells. Since we have restricted ourselves to memories, the interconnect information is made implicit by specifying the row and column locations of memory and periphery cells.

In general, circuit simulation is required to determine the extent of cell failure caused by a circuit fault [Maly 84a, Shen 85]. In order to reduce computational costs, the failure analysis module uses a set of heuristics to determine what nets and transistors have failed due to each fault. This procedure is divided into two steps: first determining what devices and nets have ceased to function properly, and then what cells contain these devices and are connected to these nets. The heuristics err on the side of conservatism. Any cell containing or using failed nets or transistors is assumed to fail. The heuristics will also report failures as propagating among several cells, when there may be only one failed cell. But the heuristics will not miss any failures.

7.2.1.1. Heuristics

The circuit fault list provided by VLASIC includes shorts, opens, shorted devices, open devices, new vias, and new devices. For the purposes of redundancy analysis, shorted devices are equivalent to a short between the transistor source and drain nets. A shorted device also causes a transistor failure, but due to the source-drain short, this is irrelevant. A new via is considered equivalent to a short between a transistor gate and the substrate.

In order to analyze short circuits, the concept of signal strength must be introduced. Nets are classified as high power, low power, or floating in order of decreasing signal strength. All nets in the cell description file must have a strength associated with them. The failure propagation rule for short circuits is that if two nets short together, the weaker net fails, while the stronger net does not. If signals of the same strength short together, both nets fail. Normally the supply rails and substrate are considered to be high power, signal nets are low power, and dynamic storage nodes are floating. If we did not have the strength designation, then a large number of global failures would occur since shorts to the supply rails are common, due to their large area. When considering short circuits, care must be taken to propagate failures correctly. For example, if a pinhole causes a storage node to be stuck at Vdd while a shorted device causes the storage node to be connected to the bit line, then the bit line is also stuck at Vdd. By suitable choice of strength designations, the short circuit heuristic will accurately reflect circuit behavior for simple circuit designs.

Open device faults cause the open device to fail, but do not affect those nets connected to it. This heuristic is realistic for digital circuits. In some circuits, the change in capacitance due to a missing channel may cause a circuit to fail. We exclude these types of faults from our analysis, since they are not DC changes to the circuit topology.

An open circuit is considered to cause the entire net to fail. This assumption is very conservative since the open may be on a branch that terminates in a cell, so only that cell will fail. This heuristic assumes that any capacitance change on the net does not affect circuit functionality.

New gate devices are ignored since they must also cause an open circuit, which is handled above. We conservatively assume that these devices will always be off. If they were on, then they might reconnect the broken nets. We also ignore new gate devices that incorporate existing transistors if their gate is floating or at ground, since they will not affect surrounding nets. Those devices that might be on are considered to cause all low power and floating nets connected to their source/drain terminals to fail. This is conservative since the pre-existing source/drain terminals of the new device will usually not cause a failure on the nets that they connect to.

New active devices are assumed to cause all nets that connect to their source/drain terminals to fail except if their gate is floating or grounded, in which case the device does not cause any failures. As with the case of new gate devices, there will be some cases where a new active device does not cause a failure, but one is reported, for example, if the new active device is in parallel with an existing device.

7.2.1.2. Cell Failures

Once we have determined what devices and nets have failed, we must determine what cells have failed. In the case of shorts, opens, shorted devices, and new devices, all those cells connected to the failed nets are assumed to fail. Open devices only cause those cells containing them to fail.

An open circuit is assumed to cause the entire net to fail. In practice, an open will only cause those cells disconnected from signal sources to fail. For example, if a clock line breaks, only those cells cut off from the clock driver fail. Automatically determining signal flow is a difficult problem [Frank 82, Jouppi 83, Ousterhout 83, Frank 85], and requires global knowledge. Most cases can be handled by detecting whether the broken net terminates in the cell, or touches several cell boundaries. The former would be considered a single-cell failure, while the latter would result in the entire net failing.

Determining what cells are connected to a net is done with the cell interconnection information. Ideally this interconnection information is part of a hierarchical wirelist that also describes the cell circuits [Frank 81, Gupta 83b, Tarolli 83, Ousterhout 84a, Ousterhout 85, Wagner 85]. The advantage of a hierarchical list is that often all those cells connected to a failed net are located in one supercell. Rather than reporting a list of failed cells, it is possible to report only the failure of the supercell. Since the supercell is often the unit of spare swapping, this form of cell failure information is more useful to the spare swapping procedure.

We use cell information that is specialized for RAMs. Each device has a cell location. Each net is located either in a bit cell, a row, a column, or other cell, and failures are reported as such. Nets in a row or column do not necessarily have to interconnect all cells in that row or column. The designation of a net as being in a row or column merely indicates that if the net fails, then the entire row or column fails. For example, nets in a row driver are designated as belonging to the row, since if they fail, the row fails. In a general-purpose failure analysis system, this design knowledge would only be available to the design-specific spare-swapping procedure. The cell information is stored in a text file and read in by the failure analysis module.

7.2.2. Spare Swapping Procedure

The spare swapping procedure must have knowledge of what cells are spare, what cells have spares, and what cells do not have spares. The structure of the spares must also be known, that is, what groups of spare cells can be swapped for what groups of failed cells, e.g. a group of spare cells in a spare row can be swapped for a group of failed cells in another row. The spare-swapping procedure must know what spare-swapping algorithm to use, and what information to record for later output.

The failure analysis procedure generates a list of failed cells for each unique chip fault list. The spare swapping procedure first determines whether any failed cells do not have spares. If so, a fatal failure has occurred, and failure analysis halts. The procedure then determines whether any of the failed cells are spare. If so, then the entire group of spare cells of which the failed spare cells are members are deducted from the available spares. For example, if a bit in a spare row fails, then that row cannot be used to repair other rows. The procedure then uses the built-in algorithm to determine how best to allocate available spares to failed cells.

We have chosen to implement a spare swapping procedure for a RAM array with row and column redundancy. Any spare row or column can be swapped for any other row or column. As discussed above, the failure analysis module categorizes failures into bits, rows, columns, and other cells rather than just providing a list of failed cells.

Failures in cells that are not part of a row or column are fatal failures since only rows and columns have spares. If there are no fatal failures, then the repair procedure first attempts to swap spare rows for failed rows and spare columns for failed columns. If there are more failed rows than spare rows, or more failed columns than spare columns, then redundancy exhaustion occurs. We assume that there are not enough spare rows to repair a failed column, and vice versa. If row and column repair succeeded then failed bits are individually repaired by first using any remaining spare rows and then columns. Rows are used before columns for bit repair because multiple bit failures in a row are more likely to occur than multiple bits in a column, based on typical layouts. Using rows first increases the likelihood that a repair will correct several failed bits with one spare element. Our repair algorithm is known to be suboptimal, but is sufficient to illustrate the feasibility of redundancy analysis.

7.3. Redundancy Analysis Examples

An example redundancy analysis was done using a 10 x 10 array of the three-transistor DRAM cell shown in Figure 6-12. The bottom two rows and left two columns are designated as spares. There are no peripheral circuits, so all cells have redundancy. The process used is the one described in Table 6-1, except that the defect density is $2000/cm^2$ rather than $20{,}000/cm^2$. A VLASIC simulation of 1000 chip samples was done, resulting in 396 unique chip fault lists. The VLASIC simulation took 25:14 minutes of CPU time and 2.1 megabytes of memory on a VAX-11/785. The fault lists were fed into RAPP, resulting in the output shown in Figure 7-2. The RAPP analysis took 12 seconds of CPU time and 680 kilobytes of memory. In general, RAPP analysis uses less than 1% of the time of the VLASIC simulation.

As noted above, no fatal failures occurred in the simulation. The Only Failed Spares line is the number of chips that only had failures in spare rows and columns. The No Failure category has a 95% confidence interval of $\pm$ 3.1%. The two spare rows and columns repaired almost all failures. The process has relatively low defect clustering, which limits the variance in the

```
SAM> rapp omcellarray.flt1 omcellarray.cell
Reading Chip Fault Lists...
Reading Design Description...
Initializing Summary...
Analyzing and Repairing Failures...
Results of Redundancy Analysis:

Total Number of Chip Samples:  1000

569 No Failure
68  Only Failed Spares
347 Repaired Failures
        66 1 Rows 1 Columns 0 Bits
        60 1 Rows 0 Columns 0 Bits
        53 0 Rows 0 Columns 1 Bits
        30 2 Rows 1 Columns 0 Bits
        25 1 Rows 0 Columns 1 Bits
        24 0 Rows 1 Columns 0 Bits
        16 1 Rows 1 Columns 1 Bits
        16 2 Rows 0 Columns 0 Bits
        14 0 Rows 0 Columns 2 Bits
        12 2 Rows 2 Columns 0 Bits
         7 1 Rows 2 Columns 0 Bits
         6 1 Rows 0 Columns 2 Bits
         5 0 Rows 2 Columns 0 Bits
         5 0 Rows 1 Columns 1 Bits
         2 1 Rows 1 Columns 2 Bits
         2 1 Rows 1 Columns 3 Bits
         1 1 Rows 2 Columns 1 Bits
         1 2 Rows 0 Columns 1 Bits
         1 0 Rows 0 Columns 3 Bits
         1 0 Rows 1 Columns 2 Bits
16  Redundancy Exhausted
         3 3 Rows 1 Columns 0 Bits
         3 3 Rows 0 Columns 0 Bits
         2 3 Rows 3 Columns 0 Bits
         2 2 Rows 3 Columns 0 Bits
         1 2 Rows 2 Columns 1 Bits
         1 3 Rows 2 Columns 0 Bits
         1 4 Rows 2 Columns 0 Bits
         1 0 Rows 3 Columns 0 Bits
         1 3 Rows 2 Columns 1 Bits
         1 1 Rows 2 Columns 2 Bits
0   Fatal Failures

Yield Improvement Factor 1.54
```

Figure 7-2: Redundancy Analysis System Output

number of faults and thus failures on a chip.

7.3.1. Comparison to Single DRAM Cell Results

A VLASIC simulation of 10,000 samples of a single DRAM cell with the above process results in a $99.6 \pm 0.1\%$ cell yield. Due to overlapping contacts and ground lines, the 10 x 10 array has 70 times the area of a single cell. A single cell is 48 x 30 microns while the array is 268 x 376 microns. If we use a simple yield analysis with negative binomial statistics, using Equation (4-16) with clustering parameter $\alpha = 100$, we find that a 10 x 10 array of cells has a yield of $77.7 +6.7 -6.1\%$, compared to $56.9 \pm 3.1\%$ for the 1000 sample array simulation. A 100,000 sample DRAM cell simulation results in a cell yield of $99.3 \pm 0.1\%$ and a predicted array yield of $66.2 +2.3 -2.2\%$. A 10,000 sample array simulation described below has a yield of $55.2 \pm 1.0\%$.

The analytic formula and simulation do not overlap, indicating that Equation (4-16) does not account for defects that land near cell boundaries in the array.

Note that the confidence interval for the 10,000 sample array simulation is completely contained in the 1000 sample simulation. The 100,000 sample DRAM simulation 95% confidence interval does not overlap the 10,000 sample DRAM interval. Even the 99.9% intervals do not overlap. The 1000 sample simulation has a 95% confidence interval of $94.1 \pm 1.5\%$ which also does not overlap the other intervals at the 95% or 99.9% level. This effect has not been observed in other yield simulations.

Table 7-1 gives the distribution of failures for the single-cell example in Chapter 6, for the same example with one-tenth the defect density, and the redundancy analysis results.

Failure	Single-Cell 20000/cm2	Single-Cell 2000/cm2	Redundancy Example
Row	36%	42%	22% (79)
Column	8	17	8 (30)
Bit	17	22	19 (68)
Row-Column-Bit	39	19	53 (194)

Table 7-1: Single-Cell Failure Distribution

In this table, the row and column failures for the single cell are those circuit faults that cause the select lines or bit line to fail. In the 10 x 10 array, there were 363 chip samples that had a fault causing a failure. The categories in the table list the fraction of those chips with just row failures, just column failures, just bit failures, and combinations of row, column, and bit failures. The numbers in parentheses are the number of chip samples in each category.

The single-cell example with lower defect density has fewer row-column-bit failures, and more single row and column failures than the higher-density case. The difference is due to the fact that at lower defect densities, row-column-bit failures are less frequent, because many row-column-bit failures are caused by multiple defects, rather than a single defect, and multiple defects are less common at lower densities. The lower density is the density that each cell in the array sees. The higher density results in the single cell receiving as many defects as the array as a whole. In addition, the array can have faults in different cells, leading to more leading to more combinations of bit, row, and column failures, and fewer row-only failures, than can occur in the single cell.

7.3.2. Analysis of Heuristics

An exact hand analysis of the 396 unique fault lists was done to determine how much error is introduced by the open circuit and new device heuristics. We make the assumption that a break in a line that terminates in a cell only causes that cell to fail, while a break in a line connecting several cells caused all cells to fail. There are 20 failures in the hand analysis that are different from the simulation. There are 5 open ground lines that go from double-column failures to bit failures, and 7 open bit lines that go from column to bit failures. These differences are due to extra poly or missing diffusion breaking active branches off of the metal ground line, which really only break the branch. There are 3 new active and 3 new gate devices that are bit failures, but are reported as column failures, and 2 new gate devices that are not failures that are reported as column failure. The analysis of these failures requires knowledge of the circuit operation. Only one of the above differences allows previously exhausted redundancy to correct all chip failures.

7.3.3. Additional Examples

A 10,000 sample RAPP analysis of the DRAM array is shown in Figure 7-3. The No Failure category has a 95% confidence interval of $\pm 1.0\%$. The 1.7% difference the 10,000 sample simulation is within the 95% confidence interval. The yield improvement factor is 3% higher in this simulation. The VLASIC simulation of 10,000 samples of the array took 232:45 minutes of CPU time and 3.9 megabytes of memory, and resulted in 2936 unique chip fault lists with a total of 2660 unique faults. The redundancy analysis took 1:42 minutes of CPU time and 3.9 megabytes of memory. Most of the RAPP

```
SAM> rapp omcellarray.flt2 omcellarray.cell
Reading Chip Fault Lists...
Reading Design Description...
Initializing Summary...
Analyzing and Repairing Failures...
Results of Redundancy Analysis:

Total Number of Chip Samples:  10000

5520 No Failure
663  Only Failed Spares
3635 Repaired Failures
      813 1 Rows 0 Columns 0 Bits
      812 1 Rows 1 Columns 0 Bits
      558 0 Rows 0 Columns 1 Bits
      212 2 Rows 1 Columns 0 Bits
      194 0 Rows 1 Columns 0 Bits
      169 1 Rows 0 Columns 1 Bits
      146 1 Rows 1 Columns 1 Bits
      134 2 Rows 0 Columns 0 Bits
      108 2 Rows 2 Columns 0 Bits
       98 0 Rows 2 Columns 0 Bits
       95 0 Rows 0 Columns 2 Bits
       75 1 Rows 2 Columns 0 Bits
       44 2 Rows 1 Columns 1 Bits
       31 0 Rows 1 Columns 1 Bits
       28 1 Rows 0 Columns 2 Bits
       25 2 Rows 0 Columns 1 Bits
       25 1 Rows 1 Columns 2 Bits
       12 0 Rows 2 Columns 1 Bits
       11 0 Rows 1 Columns 2 Bits
        8 1 Rows 2 Columns 1 Bits
        8 2 Rows 2 Columns 1 Bits
        8 0 Rows 0 Columns 3 Bits
        5 1 Rows 0 Columns 3 Bits
        4 1 Rows 1 Columns 3 Bits
        4 2 Rows 0 Columns 2 Bits
        3 0 Rows 2 Columns 2 Bits
        2 1 Rows 2 Columns 2 Bits
        1 2 Rows 1 Columns 2 Bits
        1 1 Rows 0 Columns 4 Bits
        1 2 Rows 0 Columns 3 Bits
182  Redundancy Exhausted
       30 3 Rows 2 Columns 0 Bits
       29 3 Rows 1 Columns 0 Bits
       18 3 Rows 0 Columns 0 Bits
       15 1 Rows 3 Columns 0 Bits
       12 2 Rows 3 Columns 0 Bits
       12 3 Rows 3 Columns 0 Bits
       10 2 Rows 2 Columns 1 Bits
        6 3 Rows 1 Columns 1 Bits
        6 4 Rows 2 Columns 0 Bits
        5 0 Rows 3 Columns 0 Bits
        4 3 Rows 2 Columns 1 Bits
        3 4 Rows 1 Columns 0 Bits
        3 3 Rows 4 Columns 0 Bits
        2 3 Rows 3 Columns 1 Bits
        2 3 Rows 2 Columns 2 Bits
        2 2 Rows 2 Columns 2 Bits
        2 4 Rows 3 Columns 0 Bits
        2 2 Rows 1 Columns 3 Bits
        2 4 Rows 0 Columns 0 Bits
        2 1 Rows 3 Columns 1 Bits
        2 1 Rows 2 Columns 2 Bits
        2 2 Rows 4 Columns 0 Bits
        2 2 Rows 3 Columns 1 Bits
        1 3 Rows 0 Columns 1 Bits
        1 0 Rows 4 Columns 0 Bits
        1 4 Rows 1 Columns 1 Bits
        1 0 Rows 3 Columns 1 Bits
        1 4 Rows 4 Columns 0 Bits
        1 2 Rows 2 Columns 3 Bits
        1 4 Rows 3 Columns 1 Bits
        1 1 Rows 4 Columns 0 Bits
        1 1 Rows 3 Columns 2 Bits
0    Fatal Failures

Yield Improvement Factor 1.59
```

Figure 7-3: 10,000 Sample Array Example

time and memory is used to read and store the 408 kilobyte chip fault file.

A RAPP analysis was done on a 2900 transistor register file consisting of 63 words of 8 bits each, plus one spare row and column [Walker 86b]. This register file, without redundancy, was used in the Programmable Systolic Chip [Fisher 83]. The results are shown in Figure 7-4. The No Failure category has a 95% confidence interval of $\pm 2.6\%$. The VLASIC simulation took 189:06 minutes of CPU time and 7.9 megabytes of memory, and resulted in 778 unique chip fault lists with 1776 unique faults. The register file input description took 840 kilobytes in compressed form. The redundancy analysis took 2:03 minutes of CPU time and 1.8 megabytes of memory. Most of the RAPP time and memory was used to read in and store the 260 kilobyte chip fault file.

The register file analysis illustrates the limitations of the heuristics. About 95 of the 395 fatal failures are actually repairable, or lead to redundancy exhaustion. The error is due to the conservative open circuit failure heuristic. Unlike the DRAM array, the ground lines in the register file are all on the same net, so a broken ground line results in a fatal failure, rather than just a column failure. Most of the broken ground lines are actually branches that lead to a single cell. These are repairable. However if the ground line for a column breaks, this will result in redundancy exhaustion since two columns share one ground line (except for column 8) and there is only one spare column.

```
SAM> rapp newreg.flt newreg.cell
Reading Chip Fault Lists...
Reading Design Description...
Initializing Summary...
Analyzing and Repairing Failures...
Results of Redundancy Analysis:

Total Number of Chip Samples:  1000

223 No Failure
4   Only Failed Spares
217 Repaired Failures
       94 1 Rows 0 Columns 0 Bits
       34 1 Rows 1 Columns 0 Bits
       33 0 Rows 1 Columns 0 Bits
       23 0 Rows 0 Columns 1 Bits
       13 1 Rows 0 Columns 1 Bits
       10 0 Rows 1 Columns 1 Bits
        7 0 Rows 0 Columns 2 Bits
        1 1 Rows 1 Columns 2 Bits
        1 0 Rows 1 Columns 2 Bits
        1 0 Rows 0 Columns 3 Bits
161 Redundancy Exhausted
       31 2 Rows 0 Columns 0 Bits
       20 2 Rows 1 Columns 0 Bits
       14 3 Rows 0 Columns 0 Bits
       13 0 Rows 2 Columns 0 Bits
        7 2 Rows 0 Columns 1 Bits
        5 3 Rows 1 Columns 0 Bits
        5 1 Rows 1 Columns 1 Bits
        5 4 Rows 1 Columns 0 Bits
        4 1 Rows 2 Columns 0 Bits
        4 0 Rows 2 Columns 1 Bits
        3 4 Rows 0 Columns 0 Bits
        3 0 Rows 9 Columns 0 Bits
        3 3 Rows 0 Columns 1 Bits
        3 2 Rows 0 Columns 2 Bits
        3 2 Rows 1 Columns 1 Bits
        2 2 Rows 3 Columns 1 Bits
        2 1 Rows 0 Columns 2 Bits
        2 2 Rows 2 Columns 0 Bits
        2 0 Rows 1 Columns 2 Bits
        2 3 Rows 3 Columns 0 Bits
        2 1 Rows 9 Columns 0 Bits
        2 1 Rows 1 Columns 2 Bits
        2 3 Rows 2 Columns 0 Bits
        1 0 Rows 0 Columns 3 Bits
        1 2 Rows 2 Columns 1 Bits
        1 1 Rows 1 Columns 3 Bits
        1 2 Rows 1 Columns 3 Bits
        1 2 Rows 9 Columns 0 Bits
        1 2 Rows 3 Columns 3 Bits
        1 2 Rows 3 Columns 2 Bits
        1 3 Rows 2 Columns 2 Bits
        1 1 Rows 3 Columns 1 Bits
        1 2 Rows 4 Columns 3 Bits
        1 0 Rows 3 Columns 0 Bits
        1 4 Rows 3 Columns 0 Bits
        1 0 Rows 2 Columns 2 Bits
        1 5 Rows 0 Columns 0 Bits
        1 2 Rows 3 Columns 0 Bits
        1 1 Rows 2 Columns 2 Bits
        1 4 Rows 2 Columns 0 Bits
        1 3 Rows 9 Columns 0 Bits
        1 3 Rows 3 Columns 2 Bits
        1 1 Rows 3 Columns 0 Bits
        1 0 Rows 3 Columns 1 Bits
        1 2 Rows 1 Columns 2 Bits
395 Fatal Failures

Yield Improvement Factor 1.96
```

Figure 7-4: 1000 Sample Register File Example

Chapter 8
Fabrication Data

This chapter describes the collection and use of fabrication line data for tuning VLASIC. We first describe process monitoring. We then describe a set of data taken from an operating fabrication line and its use in developing the defect statistical models described in Chapter 4. We then discuss the problems associated with the data, in particular, why it is insufficient for tuning VLASIC. We discuss how better process monitor design and sampling methods can greatly reduce the effort needed to extract the random number generator parameters.

8.1. Process Monitoring

Defect densities and sizes are determined through the use of process or defect monitors. Monitors are designed to be sensitive to one or a few defect types. Electrical measurements determine whether a monitor has failed. The number and pattern of failing monitors is most commonly used to optimize design rules and for fabrication line control. The monitors are also used to determine defect statistics. These defect statistics are used to convert from the number of failing monitors to defect densities and sizes.

The most common types of monitors are parallel plate capacitors, contact chains, interdigitated combs, and serpentine meanders, as shown in Figures 8-1, 8-2, 8-3, and 8-4. Parallel plate capacitors are used to detect oxide pinholes. Contact chains are used to detect open contacts. Interdigitated combs are used to detect intralayer short circuits. The comb fingers are wide enough to eliminate open circuits. Meanders are used to detect intralayer open circuits. The meander spacing is wide enough to eliminate short

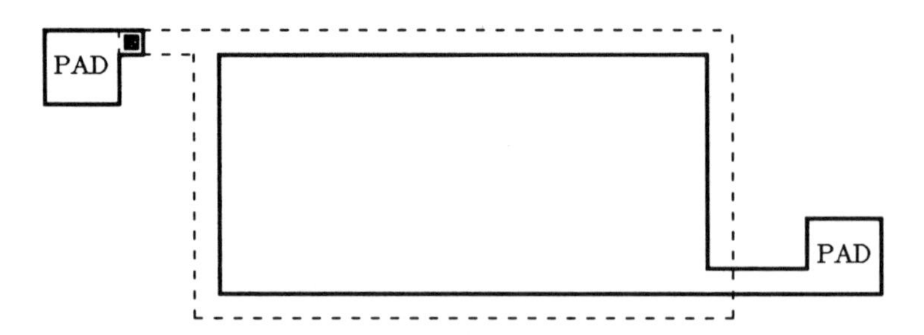

Figure 8-1: Parallel Plate Capacitor

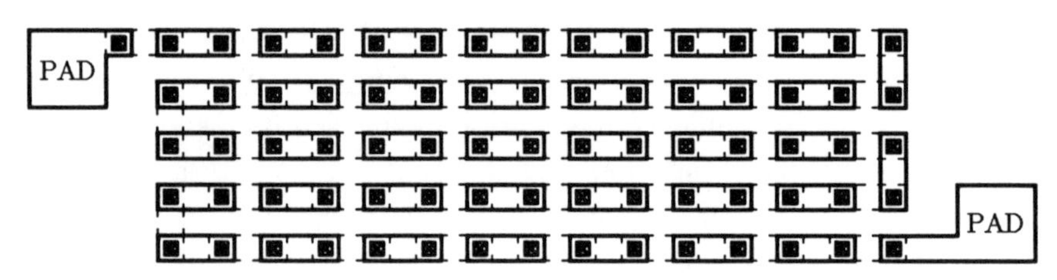

Figure 8-2: Contact Chain

Figure 8-3: Interdigitated Combs

Figure 8-4: Serpentine Meander

circuits. Combs are sometimes combined with meanders to form interdigitated meanders as shown in Figure 8-7. This allows both intralayer shorts and opens to be detected in a single monitor. Combs and/or meanders are sometimes overlapped with a perpendicular orientation as shown in Figure 8-5.

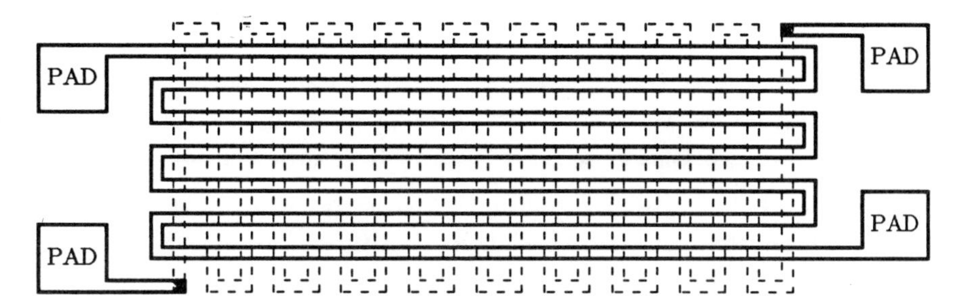

Figure 8-5: Overlapping Meanders

This allows intralayer shorts and opens, oxide pinholes, and step coverage breakage to be detected in one monitor.

8.1.1. Previous Work

Ipri and Srace used overlapping combs and meanders of different line widths, and contact strings as defect monitors for a CMOS/SOS process for design rule optimization [Ipri 77, Ipri 79]. These monitors were used to study extra and missing material defects; poly and metal step coverage; gate oxide pinholes; and open contacts. Different line widths were fabricated to study the yield versus line width. The number of good die on a wafer was maximized by determining the length of line in each layer in a typical design, and selecting optimum line widths and spacings. Defect densities and sizes were not directly computed. A similar design rule optimization was also done using an array of inverter strings [Ipri 80].

Buehler et al. developed a monitor for gate and metal-poly oxide pinholes in a CMOS process [Buehler 83]. The monitor consisted of arrays of series-parallel transistors covered by metal plates with a different number of elements (transistor plus neighboring region) in each array. After discarding failures due to global defects, the array data points were fit to Poisson statistics using the least-squares method. This provided the element failure rate for each pinhole type. The low failure rate and the global defects led to considerable spread in the data values.

Linholm et al. described a set of process monitors developed at the National Bureau of Standards (NBS) and the Jet Propulsion Laboratory of the California Institute of Technology (JPL) [Carver 80, Linholm 81, Buehler 81]. These monitors are known as the NBS test structures. These structures included an array of MOSFETs and a gate dielectric array. The latter was used for measuring gate oxide pinhole density. The MOSFET array consisted of transistors with their gate and drain connected to row lines, and source connected to columns, for random access. This array detected extra and missing epi-silicon, poly, and metal; open contacts; gate oxide pinholes; and poly and metal step coverage breaks. Visual inspection was used to determine what type of defect caused the failure.

The NBS test structures provide information about the defects causing circuit failures, but are too complex to isolate the defects to a particular process step [Mallory 83]. This requires simplified test structures sensitive to only one defect type. One such monitor is a set of combs of varying sizes, as shown in Figure 8-6. Several monitors with different comb spacings could be used to study the defect size distribution. Poisson statistics were used to derive the defect density.

Mitchell described a set of test structures for measuring extra and missing material, open contacts, and junction leakage defect densities [Mitchell 85]. Interdigitated meanders composed of a serpentine meander and interdigitated combs were used for measuring extra and missing material as shown in Figure 8-7 from Mitchell. Continuity checks between pad pairs 1-7, 2-3, 4-6, and 5-8 ensured probe contact. Extra material defects resulted in shorts between groups 1-4-6-7 and 2-3-5-8. Missing material defects resulted in opens between groups 2-3 and 5-8. Step coverage and oxide pinholes were studied by placing interdigitated meanders on different levels at right angles. Several different line widths and spacings could be used to study the defect size distribution. Open contacts were detected with contact chains of

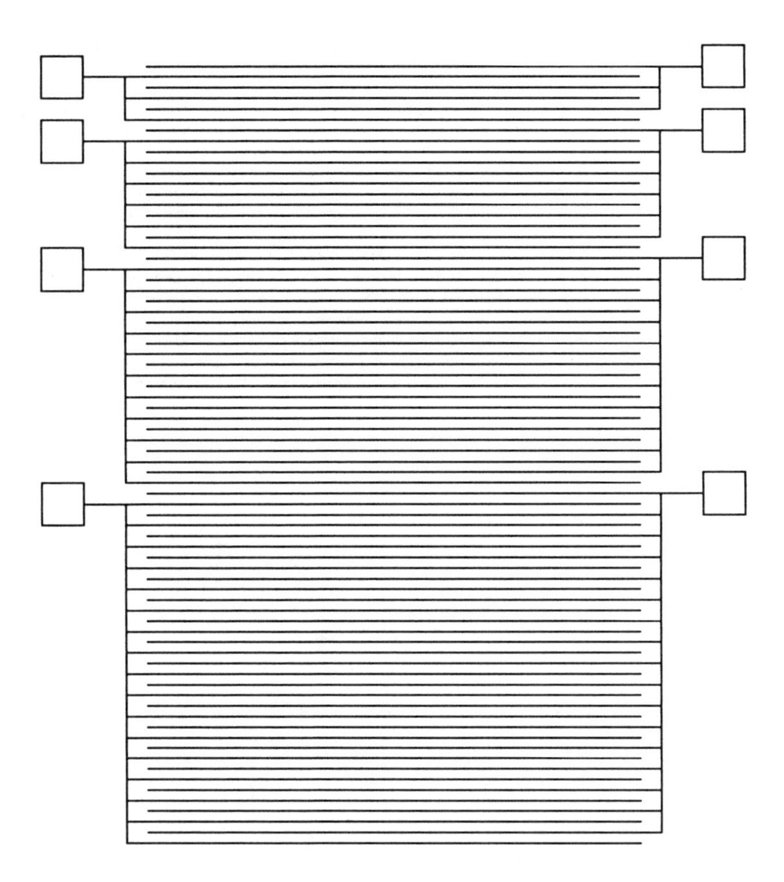

Figure 8-6: Defect Monitor for Intralayer Shorts

different lengths. Mitchell noted that when the contacts were to active, diffusion resistance and junction leakage made contact resistance tests insensitive to all but very high resistance contacts. Single-contact test structures were required to calibrate the pass/fail window. Several different monitor areas were used to allow fitting of a yield model. Negative binomial statistics were used, which required at least three monitors of different sizes.

Stapper used wafers of process monitors to detect photo defects (extra and missing material), oxide pinholes, and junction leakage [Stapper 76, Stapper 85]. The photo monitors used combs and meanders. Different widths and

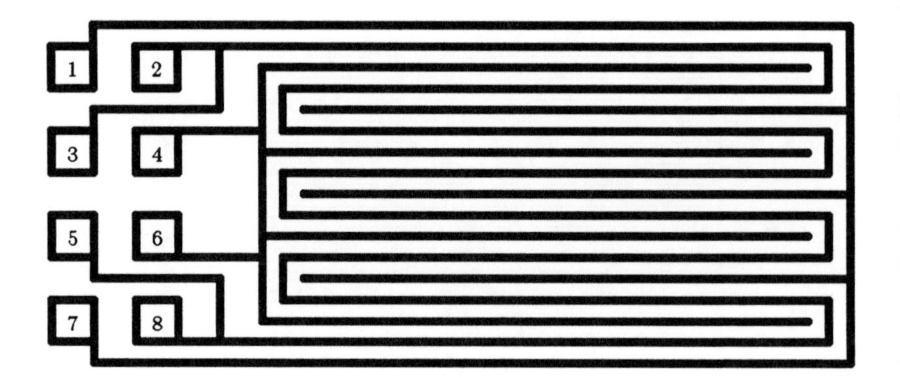

Figure 8-7: Interdigitated Meander

spacings were used to determine the defect size distribution. Both random defects and defect clusters occurred during wafer fabrication. A cluster was defined as three or more adjacent failing monitors. Clusters were removed from the sample, along with repeating defects caused by mask defects. Repeating defects are detected by examining the number of failing monitors at each location over all wafers in each lot. If the number of failures is much higher than the average, then the monitor is classified as a repeating defect, and all monitors at that location are removed from the lot data set. These clusters and repeating defects became a constant gross yield multiplier. The yield after cluster removal is the *cluster-limited* yield. Paz and Lawson modeled the gross multipier as a beta distribution [Paz 77]. The wafer minus clusters was divided into an inner and outer zones, and defect densities were estimated for each zone using negative binomial statistics. Straightforward calculation of λ and α from the mean and variance of measured values was adequate, and maximum likelihood estimators were not needed. In those cases where the variance was less than the mean, Poisson statistics was assumed [Stapper 85]. Junction leakage defects were found to exhibit strong clustering, with $\alpha = 0.4$ [Stapper 80]. Over time α has declined from 1 to 0.5.

Measuring the defect size distribution is expensive since it ordinarily requires a series of process monitors with different line widths and spacings.

Maly developed a single monitor for measuring the size distribution of extra and missing material defects [Maly 85b, Maly 87b]. Rather than using meanders and combs to detect shorts and opens, the resistance of a meander with parallel shorted layers was measured. The structure for extra and missing metal is shown in Figure 8-8 from Maly.

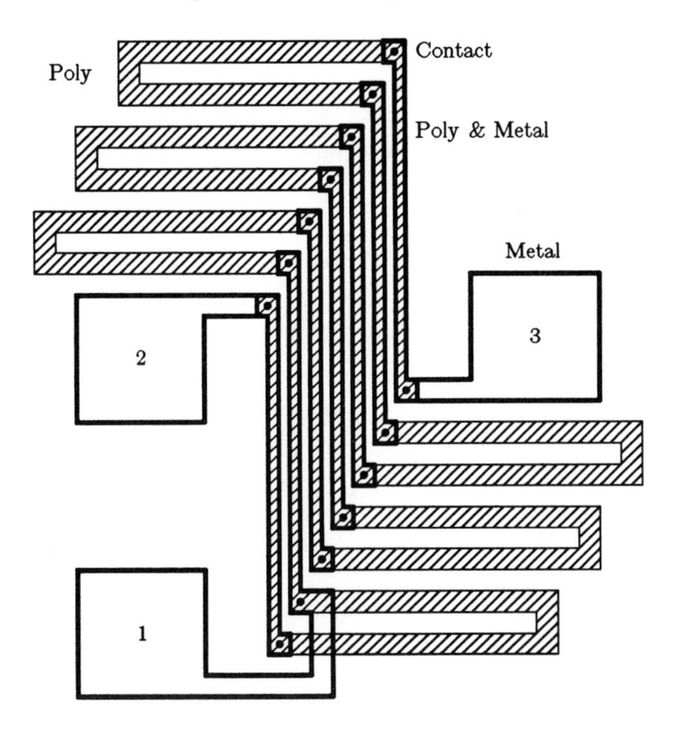

Figure 8-8: Defect Diameter Test Structure

The structure contains n (n odd) segments of poly, where a segment is either a straight vertical section or a horizontal loop. The $(n+1)/2$ vertical sections are shorted by metal. Different resistances correspond to extra and missing metal defects of different diameters. The resistance R of one segment of the poly resistor is measured between pads 2 and 3. This is used to normalize the resistance $R(n-1)/2$ between pads 1 and 3, and to make sure that no poly shorts or opens are present in the monitor. If i metal lines are shorted, then $R_{1,3} = R(n-1)/2 - iR$. If i metal lines are broken, then

$$R_{1,3} = R(n-1)/2 + iR.$$

8.2. Fabrication Data

Our fabrication data consists of 19 lots containing 278 wafers. The data was obtained from test lots taken from the fabrication line periodically over a six month period. These lots originally contained 25 to 50 wafers. Individual wafers were subjected to different experiments, so the actual number of wafers per lot with valid defect monitor data averages 15, ranging from 5 to 22. Two of the lots were incompletely tested, so we actually use 17 lots containing 214 wafers.

Each wafer contains 42 steps of a three-part die. The first section of the die contains a 16 K-bit SRAM. The second section of the die contains contact chains, parallel plate capacitors, and parametric process monitors. The contact chains test metal1-metal2, metal1-poly, and metal1-active vias. The parallel plate capacitors test the gate, first intermediate, and second intermediate oxides. The second section also contains small meanders that we do not use. The third section of the die contains three large interdigitated meanders overlapped by a meander on another layer. The first structure contains an interdigitated metal1 meander and a metal2 serpentine meander. The metal1 is placed directly on top of coincident poly lines in order to increase the step size for metal2. The second structure contains an interdigitated active meander. The third structure contains an interdigitated poly meander in an active region and a metal1 serpentine meander. There were no monitors for measuring junction leakage defects or metal2 shorts. Only 39 steps of the third part were valid. These large monitors were designed to have approximately the same statistical characteristics (line length, spacing, etc.) as the SRAM.

There are 8982 measurements from the second part of the die, and 8034 measurements from the third part of the die. These values are less than the 8988 and 8346 expected because of test failures and discarded tests due to process experiments such as varying poly width. We have kept those wafers where the experiments did not deviate too far from nominal process conditions. To discard all such wafers would make the data set too small.

The monitor test procedure was to first make an electrical measurement by forcing a current through the structure and measuring the voltage. If the monitor fell outside the voltage acceptance window, then it was visually inspected to confirm the presence of a short or open. In some cases, bad process conditions caused a large number of monitors on a wafer to fall just outside the acceptance window even though no defect was present. These wafers were detected by examining the raw probe data voltage values, and excluded from the model-fitting described below. Each monitor is used for detecting shorts and opens on two layers, and shorts between the two layers. Testing proceeded by first checking for interlayer shorts. If a short existed, no more testing was done on the monitor. Similarly for the interdigitated meanders, meander continuity was checked before measuring shorts between the meander and combs, and the second test was not done if the first failed. This led to missing data points.

8.3. Model Fitting

As described in Chapter 4 and Section 8.1.1, researchers have noted a defect size distribution, and a defect spatial distribution, with a radial distribution across the wafer, and defect clustering between wafers. We examined our data to determine what statistical models fit best and derive the composite statistical model shown in Figure 4-7. We used a combination of hand analysis and process monitor data analysis tools totally 3500 lines of code.

8.3.1. Defect Size Distribution

Our combs and meanders do not have varying line width and spacing, so it is not possible to measure the defect size distribution. However defect density measurements from monitors with different bloats and from processes in the same facility with different line widths indicates that extra and missing material defects obey Stapper's $1/x^3$ size distribution. We assume that the diameter of peak frequency x_0 is at the resolution limit of the lithography process, which is much smaller than the minimum design rule.

8.3.2. Radial Distribution

Wafer maps clearly showed that process monitors towards the wafer edge were much more likely to fail than those near the wafer center for most defect types. As described in Section 4.2.2, other researchers have observed a radial distribution in defect density. We performed a least-squares fit of the following radial distributions to the data:

$$h(r) = c_1 r^2 + c_2 r + c_3 \quad \text{Second-order polynomial} \tag{8-1}$$

$$h(r) = c_1 r^2 + c_2 \quad \text{Parabola} \tag{8-2}$$

$$h(r) = c_1 r + c_2 \quad \text{Line} \tag{8-3}$$

$$h(r) = c_1 \quad \text{Constant} \tag{8-4}$$

$$h(r) = c_1 e^{c_2 r} \quad \text{Two-variable exponential} \tag{8-5}$$

$$h(r) = c_1 + c_2 e^{c_3 r} \quad \text{Three-variable exponential} \tag{8-6}$$

$$h(r) = \begin{cases} c_1, & r \leq c_3 \\ c_1 + c_2(r - c_3), & r \geq c_3 \end{cases} \quad \text{Piecewise linear} \tag{8-7}$$

$$h(r) = \begin{cases} c_1, & r < c_3 \\ c_2, & r \geq c_3 \end{cases} \quad \text{Two zones} \tag{8-8}$$

All of the radial distributions are rotationally symmetric and obey the constraint

$$\int_0^R h(r)dr = 1. \tag{8-9}$$

Equation (8-9) constrains $h(r)$ to represent the relative defect density at a particular radius.

We first fit the functions to the monitor failure rates for all wafers in all lots combined. The two-zone distribution shown in Figure 4-6 provided the best fit to the data for 13 of 20 defect types, with a least squares sum 0.2% to 15% smaller than other distributions. In 4 of the 7 remaining cases, the piecewise linear model fit 2% to 18% better. In two of the remaining three cases, the three-variable exponential fit 0.3% and 12% better. In the final case, the second-degree polynomial fit 1% better. In those cases where the two-zone model was not the best, the monitor failure rate was sharply higher for those monitors at the edge of the wafer. This led to extremely steep slopes in the piecewise linear and exponential models. We felt that these extreme slopes were not valid, being based on so few wafers and a coarse sampling across the wafer. The slope was such that the defect density would vary by orders of magnitude across the process monitor at the wafer edge. We selected the two-zone radial distribution model as being most parsimonious with the available data.

The two-zone model divides the wafer into a circular inner zone and concentric outer zone with different mean defect densities. Defects are assumed to be randomly distributed within a zone. The two-zone distribution is shown in Figure 4-6. The inner zone radius ranged from 1.1 to 4.1 cm on 10 cm diameter wafers. The relative defect densities between outer and inner zone ranged from 0.38:1 to 3.95:1. The usual assumption that defects are always more frequent towards the edge of the wafer is not valid for this process. All the lithography process steps had higher defect densities towards the edge of the wafer except for poly opens and active shorts. The oxidation steps had higher defect densities towards the center of the wafer except for the first intermediate oxide. This may be explained by the possibility that the radial distribution is caused primarily by gross defects created during wafer handling, rather than by random defects [Ferris-Prabhu 87].

During the fitting process, it was noted that a test on two monitor sites had a near-100% failure rate in two lots. These monitors were located in the center of the wafer in lots that were processed a few days apart with the same mask set. These tests most likely failed due to repeating mask defects, and were discarded from the data set. These repeaters were detected by examining all monitor sites in each lot that had tests with a failure rate significantly higher than the surrounding monitors. In this particular case, the failure rate was sharply higher, and so easy to detect.

As mentioned previously, some researchers have defined adjacent failing monitors as a defect cluster. We felt that our monitor placement was too coarse (39-42 monitors per wafer) to justify placing monitors into clusters. In addition, except for monitors at the wafer edge, significant clustering was not observed.

We also examined whether different defect types were independent, so that a defect occurring during one process step does not affect the probability of a defect occurring during a later step. We examined wafer maps, and found that there was no correlation between defects on different layers. This supports the defect independence assumption made in Chapter 4.

We also examined fitting the two-zone radial distribution on a per-lot basis. We found that the zone radius and proportions varied tremendously from lot to lot. One problem with this fitting is that some lots had no failing monitors for a test, so a distribution could not be fit. In addition, some lots had as few as five wafers, and the fitted outer zone often contained only three or four die locations. We felt that this per-lot fitting was not valid. For full 50-wafer lots, a per-lot radial distribution might be fit, however no such attempts have been reported in the literature.

8.3.3. Negative Binomial Distribution

The data clearly indicated that the variance in the number of failing monitors per wafer was much higher than the mean. This suggested the use of a negative binomial distribution to determine the number of defects per wafer. We first attempted to fit a single negative binomial distribution for all wafers across all lots. This fitting succeeded for only a few defect types with low defect densities. It was observed that most of the variance in the number of failing monitors was between lots. An F-test on the mean number of defects per lot was performed with results shown in Table 8-1. In some

Test	Wafer Count N	F(16,N-17)
Metal1-Metal2 Contact Open	214	32.96 * 1
Metal1-Poly Contact Open	214	1.47
Metal1-Active Contact Open	214	1.12
Active-Poly Contact Open	214	2.65 *
Metal2-Metal1 Short 1	214	12.52 *
Metal2-Metal1 Short 2	208	2.54 *
Metal2-Metal1 Short 3	206	13.64 *
Metal1-Poly Short 1	214	1.98 *
Metal1-Poly Short 2	206	10.26 *
Poly-Active Short	206	31.37 *
Metal2-Metal2 Short	208	84.79 * 1
Metal2 Open 1	214	19.75 * 2
Metal2 Open 2	206	25.74 * 2
Metal1-Metal1 Short	206	19.96 *
Metal1 Open 1	206	2.51 *
Metal1 Open 2	206	1.85 **
Poly-Poly Short	205	2.51 *
Poly Open	205	3.64 *
Active-Active Short	206	19.08 *
Active Open	206	3.05 *

* Significant at 97.5% level
** Significant at 95% level
1 Most monitors failed in one lot
2 Most monitors failed in two lots

Table 8-1: F-Test for Lot Means

cases there are several different test structures measuring the same defect type. All but two test structures are significant at the 95% level, indicating that the lots have different mean defect counts.

There are only 17 lots of valid data, so a between-lot distribution could not be fit with any validity. We assume that a negative binomial is appropriate due to the fact that the observed variance is much greater than the mean.

Within a lot, the variance in the number of defects between wafers was much smaller, but usually larger than the mean. We successfully fit a negative binomial distribution within each lot. As mentioned above, most of the monitors in some lots were dead, so they were not included in the analysis. In those lots where the variance was less than the mean, we assumed a Poisson distribution, with $\alpha = \infty$. We used the method of moments for calculating λ and α as described in Equations (4-14) and (4-15).

We found that these values were always within 10% of the maximum likelihood estimators, and much easier to calculate. Stapper made a similar observation [Stapper 85].

The negative binomial clustering coefficient α for the between-wafer distribution was observed to vary between lots, ranging from 0.06 to ∞ for a single defect type. In the most extreme case, some lots had no failing monitors, while others had uniformly distributed failing monitors. Due to the small number of wafers within the test lots, one or two wafers with many failing monitors could strongly influence the clustering coefficient. Stapper noted the same phenomenon, but found that it was due to sample limitations since long term averages showed constant α values [Stapper 85]. Like him, we assume that the clustering coefficient for wafers within lots is the same for all lots. We use the average α value for all lots.

Rather than considering whole wafers, Stapper considered inner and outer wafer zones separately when fitting a negative binomial distribution across all wafers [Stapper 76]. We attempted to fit a negative binomial distribution within zones for all lots, and within zones for each lot separately. This effort fared poorly. The outer zone usually had only 5-10 defect monitors in it, so there were only a few monitors in the outer zone per lot. The fitting was not nearly as good as when whole wafers were used.

In order to translate from the number of failing monitors to defect densities, we use the assumed defect size distribution to estimate the effective critical area. We then use this critical area to determine the mean defect density D using the relation $\lambda = AD$ where λ is the expected number of defects, A is the critical area, and there is no within-monitor defect clustering. Ferris-Prabhu noted the difficulty in attempting to simultaneously extract both the spatial and size distribution from the process monitors [Ferris-Prabhu 85c].

Even though we could not tune the distributions to the process, it is possible to determine a set of realistic parameter values for a three-micron process analogous to the synthetic process conditions given in Tables 6-1 and

6-3. This realistic process is shown in Table 8-2. These process conditions should result in yields of 10 to 50% for one-cm^2 chips. Note that the defect densities are about 1000 times lower than those given in the synthetic processes.

Defect Density	
Extra/Missing Metal1	20/cm2
Extra/Missing Metal2	20/cm2
Extra/Missing Poly	20/cm2
Extra/Missing Active	20/cm2
First-Level Pinholes	10/cm2
Second-Level Pinholes	10/cm2
Gate Oxide Pinholes	10/cm2
Design Rules	
Metal1 Width/Space	4.5u
Metal1 Contact Width	3.0u
Metal2 Width/Space	4.5u
Metal2 Contact Width	3.0u
Poly Width/Space	3.0u
Active Width/Space	3.0u
Poly/Active Space	1.5u
Diameter of Peak Density	
Extra/Missing Metal	0.5u
Extra/Missing Poly	0.5u
Extra/Missing Active	0.5u
Maximum Defect Diameter	18u
Between Lot Alpha	0.5
Between Wafer Alpha	100
Wafers Per Lot	50
Wafer Diameter	12.5cm
Radial Distribution	
Zone Diameter	4.5cm
Inner Zone Fraction	0.3
Outer Zone Fraction	0.7
Minimum Line Spacing	0.5u
Minimum Line Width	0.5u

Table 8-2: Realistic Process Conditions

8.4. Fabrication Data Problems

The data described in Section 8.2 allowed us to select models for some but not all defect distributions. We could not fit a defect size distribution or a between-lot spatial distribution. The data only provides indications that the model used in Chapter 4 is correct. The data can not be used to tune the models to the fabrication line. The data suffers from three basic problems:

insufficient data, missing data, and data contamination.

8.4.1. Insufficient Data

Our data consists of 8988 die on 214 wafers in 17 lots. This is substantially more than is used in most of the yield modeling efforts in the literature. This data was gathered over a six month period at considerable expense. Despite this fact, the data was insufficient to fit all defect distributions or tune the distributions that can be fit.

The most glaring deficiency is the lack of defect size data. Several interdigitated meander structures of different line widths and spacings are required to estimate the size distribution. Stapper used six open circuit and six short circuit monitors with the width and spacing varying over a six to one range [Stapper 84a]. However this would multiply the number of monitors required by a factor of twelve, which is too great. One solution would be to use Maly's structure described in Section 8.1. The disadvantage of this structure is that it has a resolution of $W+S$ for width and spacing W and S and an uncertainty in defect diameter of $\pm(W+S)$. This is sufficient for large defects, but inadequate for small defects. The problem can be solved by supplementing the monitor with a few combs and meanders with narrow widths and spacings.

The large size of the meander and comb monitors, and the presence of the SRAM and parametric test structures means that only a few (39 to 42) monitors can be placed on a wafer. This makes it difficult to fit a radial distribution, and impossible to study intrawafer defect clustering. Discarding the SRAM and other test structures would allow three times as many monitors to be placed on the wafer. In addition, the monitors could be made much smaller.

The number of wafers per test lot is often too small (as few as 5) to tune a between-wafer negative binomial distribution. The primary reason for this is experimentation on many of the wafers. If all 25 to 50 wafers per lot were fabricated using nominal process conditions, then sufficient data would be available for fitting the between-wafer distribution.

The number of test lots is much too small for fitting a between-lot spatial distribution. The problem is that test lots are expensive. One hundred test lots can cost several million dollars on a state-of-the-art fab line. In the past an alternative was to use process monitor drop-ins on production wafers. Use of step-and-repeat lithography makes this option very undesirable. One way to put process monitors on production wafers is to put the monitors in the scribe lanes. This technique is used for parametric monitors. Most catastrophic monitors are too large to fit in the scribe lanes, but have been used on occasion [Stapper 76]. A compromise between test lots and test patterns on every wafer is to include one or two test wafers in each lot. The data provided from these wafers, supplemented by test lots, should be sufficient to predict the between-lot defect distribution.

As noted previously, we do not have any data on metal2 shorts, and assume that they obey the same statistics and have the same defect density as metal1. There is also no data for junction leakage defects, however they appear to be rare enough in the SRAM that they can be neglected when analyzing static circuits. Data on metal2 shorts can be obtained by modifying the test structures so that the metal2 serpentine meander is replaced with an interdigitated meander. Junction leakage can be measured using the active interdigitated meander.

8.4.2. Missing Data

As noted above, the process monitors are overloaded, in that they are used to detect several different defect types. The metal1 interdigitated meander covered by a metal2 serpentine meander is used to detect metal1 shorts, metal1 opens, metal2 opens, and metal1-metal2 shorts. The test order was to first check for metal1-metal2 shorts, then metal2 continuity, then metal1 meander continuity, and then metal1 shorts. Testing stopped on the first failure, leading to missing data points.

The missing data is biased. Monitors with one type of failure are likely to be located in those regions of the wafer with above-average failure rates for other defect types. Thus the missing data is biased towards failure. Ignoring the missing data, as we did in Section 8.3, leads to predicted defect densities lower than the actual values. There are some excepts to this rule. Metal1-metal2 shorts are more likely to occur towards the center of the wafer, while metal1 and metal2 opens and intralayer shorts are more likely to occur towards the edge of the wafer.

One way to get around the missing data problem is to estimate the monitor failure rate from the zone failure rate, and then to synthesize values for the missing monitors. This problem of missing biased data is similar to the problem faced by the U.S. Census Bureau. People of lower socio-economic status are less likely to be counted. This problem is an area of current statistical research.

In most cases, the missing biased data problem can be eliminated by a change in test procedure. There is no need to halt testing at the first failure. For example, if there is a metal1-metal2 short, it is still possible to perform all other measurements on the monitor. Even if a meander is open, it is still possible to measure shorts since there is a probe pad at each end of the meander, and the open will affect only a small area.

Missing data points also occur due to probe failure. Some of the probe pads were extremely close to the wafer edge, and electrical contact could rarely be made to them. These data points were excluded from the analysis since these failures are not random in nature. Probe failure occasionally happened on other parts of the wafer, but this occurred randomly, and was rare enough to neglect.

Repeating defects due to mask defects also lead to missing data points, since all monitor tests in the affected lot must be discarded. There were only two instances of repeating defects in the data set. These do not significantly affect the analysis. Processes that use 10:1 step-and-repeat lithography do not suffer as much from repeating defects.

Two of the test lots were discarded due to major deviation from nominal parameters. These lots undoubtedly have very high defect densities, however it is not meaningful to include them in our analysis since we are interested in simulating local defects in a fabrication line near its nominal operating conditions.

8.4.3. Data Contamination

The process monitors are affected by parametric as well as catastrophic defects. As was noted in Chapter 1, these two defect classes are coupled. Line width variation affects the critical area of the meanders and combs to shorts and opens. This effect can be accounted for by measuring the line width and adjusting A when calculating the defect density D with the formula $\lambda = AD$. We do not have this line width information directly available (although van der Pauw and resistor structures are present), so the data used for the model fitting was subject to these variations. As noted above, those monitors that deviated too far from nominal conditions were excluded from the analysis.

The data is also contaminated with local defects that are not modeled by VLASIC. Hillocks on the metal1 layer cause metal1-metal2 shorts. Hillocks are more likely to form on large metal1 regions, so large metal1-metal2 parallel plate capacitors are more likely to fail than a capacitor of equal area composed of many small regions. This problem can be solved by measuring metal1-metal2 shorts with perpendicular meanders. This data is available in the second section of the test chip, and is used instead of the parallel plate capacitor data. VLASIC does not handle step coverage breakage. The test structures in section two of the test die were designed to provide large steps for metal1 and metal2. Visual observation showed good step coverage, so this type of defect can be ignored for this process.

8.5. Economical Process Monitoring

The model fitting procedure described in Section 8.3 required man-months of effort, and was not complete. A high-quality model fitting and tuning may require six months or more of work. This is too expensive in an environment where processes change every few years. We would like to have tools that allow the initial model selection and tuning to be done within a few days. The problem of tuning VLASIC is similar to the problem of tuning FABRICS with Prometheus [Spanos 83, Spanos 85a, Spanos 85b], except in the catastrophic, rather than parametric defect domain.

Once we have selected defect distributions and performed an initial tuning, we would like to keep them tuned as process conditions change from week to week and month to month. This requires large amounts of data to be collected. Scribe lane test patterns and test wafers may not be sufficient for the task. A partial solution is to use optical techniques to detect particles and lithography defects without process monitors. Particles can be discovered with an oblique bright light [Stapper 82b] or a scanning laser beam [de la Rosa 86]. A scanning electron microscope could be used for detecting smaller particles. Lithography defects can be detected by comparing the fabricated pattern with the desired pattern. This comparison must be done immediately after the lithography step before additional layers have been deposited. The cost of these techniques is the direct cost of the measurement and the additional inventory cost of wafers undergoing test. The advantage of these techniques is that they provide real-time data. In the future this data may be used for interactive fab line control. These techniques may have difficulty measuring the defect diameter distribution. Optical techniques cannot be used to detect material defects such as junction leakage and oxide pinholes. However these types of defects can be measured with relatively small monitors and so use scribe lane monitoring.

One possible solution to the data gathering problem is to use the product itself as a defect monitor. Gangatirkar et al. used the bit maps generated by testing DRAMs for process diagnosis [Gangatirkar 82]. Bit failures were classified into several patterns, as shown in Figure 8-9. A critical area

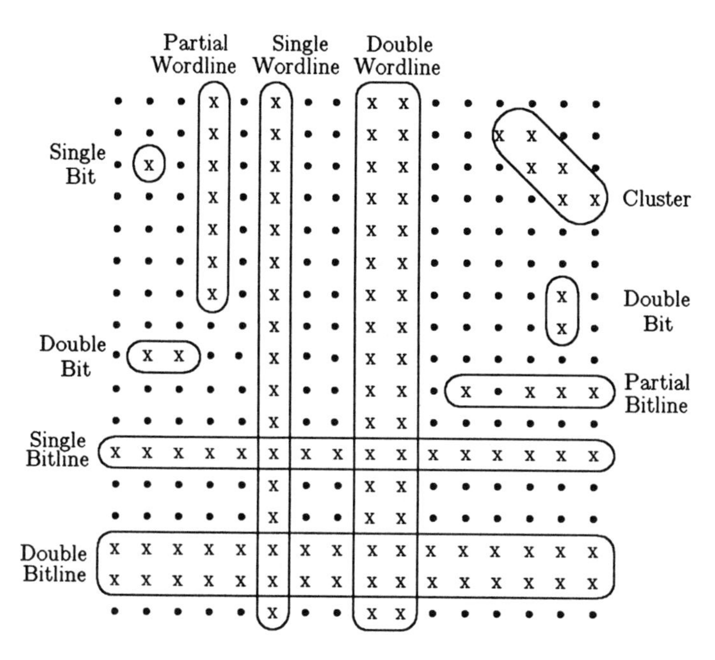

Figure 8-9: Some Typical Bit Failure Patterns

analysis was done using the known defect size distribution to determine the probability that a defect causes a failure pattern, as shown in Table 8-3.

Failure Patterns	Process Defect Type -Metal1	+Metal1	-Poly	+Poly	Pattern Count
Single Bit	0.3	0.4	0.0	0.0	170
Double Bit	0.2	0.1	0.0	0.0	80
Cluster	0.1	0.2	0.0	0.0	70
Partial Wordline	0.0	0.0	0.0	0.4	320
Single Wordline	0.1	0.1	0.5	0.1	330
Double Wordline	0.0	0.1	0.0	0.0	20
Partial Bitline	0.1	0.0	0.2	0.1	190
Single Bitline	0.1	0.0	0.1	0.1	150
Double Bitline	0.0	0.0	0.1	0.0	40
All Others	0.1	0.1	0.1	0.3	330
Defect Count	300	200	400	800	

Table 8-3: Failure Pattern Probability Matrix

Using the observed failure patterns and the probability matrix, individual defect densities were estimated using negative binomial statistics. This is essentially the reverse of the yield calculation carried out in [Stapper 80].

Related techniques have been used for diagnosing parametric defects [Strojwas 82, Strojwas 85, Odryna 85a, Odryna 85b] and are being developed for identifying groups of process steps with high spot defect density.

8.5.1. Test Analysis Post-Processor

We are developing a system called TAPP (Test Analysis Post-Processor) to generate failure pattern probability matrices from VLASIC simulations. For memory applications the system performs adequately. However for general circuits, the matrices become very large and sparse. There are several reasons for this. First, failure patterns are actually combinations of failing test patterns. For memories, these can be reduced to a few categories. However in the general case, there can be thousands or millions of such combinations generated by the tester. Second, the defect categories are not single defect types, but are actually defect combinations. Normally defects are rare enough that they can be considered independently. However during pilot production, when the yield is low, defects may interact, and all the defects on a die must be considered together. It is also desirable to use the test data to extract the defect size distribution as well as the defect density. This can be done by further dividing the defect categories by defect size range. This only makes the problem of many defect categories worse.

There are three additional considerations in building the probability matrix: global process disturbances, masking failures, and tester limitations. Global process disturbances such as line width variation and mask misregistration modify the entries in the probability matrix. For example, poly line widths smaller than the nominal value will increase the probability that a missing poly defect will cause a wordline break in a RAM. When analyzing test data, parametric process monitors must first be measured to determine the global process variations. The appropriate probability matrix can then be selected. Generating the matrix is too expensive to do for each set of measured global process disturbances. Instead matrices are generated for several different global process disturbances, and then interpolation is used to approximate the correct matrix.

As mentioned previously, when the defect density is high, defects can interact. One type of interaction is masking failure. One defect may cause a failure that masks the presence of another failure. For example, a missing metal defect can break a bitline. Then all other defects causing cell failures in that bitline are masked. This becomes a problem when one frequently-occurring defect type tends to mask another frequently-occurring defect type. The result is that the second defect type will be undercounted. When the extracted defect density is applied to a different chip design, the predicted yield will be incorrect. In order to make correct yield predictions, the estimated defect densities must be adjusted to account for masking. The extent of masking can be determined from VLASIC simulations.

Ideally test data analysis should be performed using existing test programs, so that the analysis is nearly zero cost. However this introduces a number of difficulties. Test programs are normally designed so that they reject the part when the first failure is encountered. In fact one application of the failure probabilities is to minimize the average test time by testing for the most likely failures first [Maly 84a, Shen 85, Maly 87a]. The result is that, particularly in low-yield processes, most failures are not detected, since the chip has already been rejected. For a given test sequence, it can be determined the order in which failures will be detected. Then each failure masks all failures that would have been detected later. As is the case for masking failures, the estimated defect densities can then be adjusted to more closely reflect their true values.

Chapter 9
Conclusions and Current Research

This chapter describes our conclusions and directions of current research. We first discuss what we have learned in the course of this research. We then discuss how this research has contributed to the field of yield modeling and simulation. We then discuss how this work is being extended to improve simulator performance, use new analysis techniques, and how local and global defects can both be combined in one simulator.

9.1. Conclusions

VLASIC automates the process of simulating circuit faults due to catastrophic local defects. Current practice is to repeatedly place defects on a large checkplot by hand and visually determine whether a fault has occurred. Whereas hand simulation involves thousands of defects, VLASIC allows the placement and analysis of hundreds of thousands or millions of defects.

VLASIC also provides a framework for research on the simulation of local defects. The table-driven defect models are readily extended to other MOS processes. For example, in a CMOS/SOS process, extra and missing epi-silicon can be modeled as circles of extra and missing material [Ipri 77, Ipri 79, Linholm 81]. Some modifications to the models will be necessary to handle new types of defects. For example, in metal-gate CMOS/SOS, gate oxide pinholes primarily occur at the silicon/sapphire interface [Bernard 78]. The defect models currently assume that the number of oxide pinholes is a function of oxide area. Most defects unique to bipolar processes, such as diffusion pipes, can be handled within our existing framework. Pipes can be modeled as points that cause an emitter-collector short when they intersect the emitter region.

The defects and their models used in this research are listed in Table 9-1.

Defect Type	Model
Extra/Missing First Metal	Circle
Extra/Missing Second Metal	Circle
Extra/Missing Polysilicon	Circle
Extra/Missing Active	Circle
Extra/Missing First Vias	Circle
Extra/Missing Second Vias	Circle
First-Level Oxide Pinholes	Point
Second-Level Oxide Pinholes	Point
Gate Oxide Pinholes	Point
Junction Leakage	Point

Table 9-1: Defect Types and Models

The defect models that we describe in Chapter 3 have only been touched upon in the literature. Our models provide a uniform way of treating defects and the circuit faults that result from them.

In Chapter 4 we describe a composite model for defect statistics. This model is composed of distributions taken from the literature, but that have never before been combined in our manner. Our composite distribution also models effects such as between-lot defect density variance that have not been discussed in the literature.

In Chapter 5 we show that type-driven fault analysis provides the most efficient method of applying the defect models in order to determine what faults result from defects placed on the layout. We indicated the problems associated with other fault analysis techniques. This provides impetus for current research, as discussed below.

We have chosen to implement VLASIC as a Monte Carlo simulator, as described in Chapter 6. This approach allows us to make a smooth tradeoff between execution time and simulation accuracy. VLASIC can be used to quantify errors introduced by approximation techniques. The Monte Carlo simulator provides a framework for current research through its modular organization. Defect models, defect statistics, fault analysis procedures, and application post-processors can be modified or replaced without disturbing the rest of the system.

Our redundancy analysis system discussed in Chapter 7 confirms the usefulness of our approach to yield simulation. We show how yield simulation provides a more accurate determination of yield in the presence of redundancy than simple yield formulas. We also gain insight into what design tradeoffs to make when selecting the level and type of redundancy.

In our examination of fabrication data in Chapter 8, we show that our statistical models fit the data reasonably well. We also show that the types of information needed for yield simulation are quite different from process control. Our work provides feedback for selecting better process monitor designs and sampling methods. Unfortunately our fabrication data was insufficient to tune the random number generators, so validation of the simulator could not be performed. However related yield analysis work provides strong indications that our approach is valid.

In conclusion, VLASIC represents a first step at modeling and simulating local defects. This work will lead to the development of a body of knowledge about local defects similar to what already exists for global defects. As discussed below, these two veins of research must eventually merge.

9.2. Current Research

Current research in yield simulation focuses on three areas: increasing the performance of the Monte Carlo simulator, using non-Monte Carlo analysis techniques, and combining local and global defects into one simulator.

9.2.1. Faster Monte Carlo Simulation

VLASIC is not as fast as was originally hoped. Large simulations take about 30 msec per placed defect, and 80-150 msec per chip sample. There is also an initialization time of about 350 msec per transistor, or 10 hours for a

100,000 transistor chip (a 16 K-bit SRAM). A simulation of 100 boats of 50 wafers each with 100 die per wafer (500,000 samples) takes 21 to 31 hours of CPU time. The Monte Carlo nature of VLASIC means that it is in a class of programs termed "embarrassingly parallel." Multiprocessors can be used to greatly reduce VLASIC execution times. Part of VLASIC's low performance is due to a time-space tradeoff. A 16 K-bit SRAM layout contains more than two million rectangles. The VLASIC bin arrays must be kept as small as possible to allow problems to fit into a reasonable amount of memory (20 megabytes). The hierarchical design specification of the SRAM contains only a few thousand rectangles, since the layout is built from regular structures. Storing this hierarchical description greatly reduces space needs. We first discuss parallel VLASIC implementations. We then discuss hierarchical implementations and how this additional space can be used for performance improvement. We then discuss how simple but fast fault analysis might be done.

9.2.1.1. Parallel Implementations

Monte Carlo simulators like VLASIC can be easily transported to a shared-memory multiprocessor. There are several levels of potential parallelism. The most straightforward implementation uses *sample* parallelism. The intermediate layout structures are created serially. Then each of m processors places and analyzes the defects for n/m of the n chip samples, saving the results in a list. At the end of the computation, these lists are combined by totaling the counts for each fault group combination. Sample parallelism is not identical to a uniprocessor simulation since m random number generators are used. The parallel result can be obtained on a uniprocessor by using each of the m random number generators in turn. Sample parallelism provides linear speedup for the sample generation phase of the simulation. Only a small percentage of the time is spent accessing the shared layout data structures, so caches can be used effectively, and interconnect bandwidth requirements are not high.

A further speedup can be gained by placing and analyzing all the defects on a chip sample in parallel. This is called *trial* parallelism. The next level of speedup would be to perform defect generation, placement, and analysis in parallel. The defect analysis phase can be sped up by performing some individual polygon operations in parallel. For large chips, a large amount of time is spent building the intermediate data structures. Most of this time is spent performing local polygon operations. Since these operations only affect local neighborhoods of the layout, they can be performed in parallel by partitioning up the layout. It is possible to break polygon operations down into individual edge operations, however the grain size would be too small to amortize the task scheduling overhead. Since a typical simulation involves thousands or tens of thousands of chip samples, with tens of defect trials per sample, a VLASIC simulation will show linear speedup for thousands of processors using only sample and trial parallelism.

A multicomputer (non-shared memory multiprocessor) implementation of VLASIC is substantially more difficult. For large chips, the intermediate data structures are too large to fit in private memory. The layout must be divided up among the processors. This makes it natural to generate the intermediate data structures in parallel from the partitioned layout. Multicomputers generally have slow serial links, so it is desirable to analyze each defect in the processor that contains the layout neighborhood using trial parallelism. One processor generates the chip samples, and distributes the trial tasks according to the defect location. The communication overhead for each task is low. Since the defects are randomly distributed across the chip, the work will be evenly distributed among the processors. The sample generation can be done by several processors as before, with the fault lists being combined at the end of the simulation. Additional levels of parallelism can be used as described above. As was the case for the shared-memory architecture, linear speedup will occur for thousands of processors.

9.2.1.2. Hierarchical Implementations

General hierarchical techniques for integrated circuit analysis have been studied by several researchers [Hon 83, Scheffer 81, Tucker 82, Newell 82, Losleben 79]. Most hierarchical analysis has concentrated on design rule checking [McGrath 80, Whitney 81, Marek-Sadowska 82, Maly 82b, Taylor

84], and circuit extraction [Gupta 83b], or both [Wagner 85, Ousterhout 84a, Ousterhout 85]. In some cases, restrictions were placed on the layout hierarchy in order to simplify analysis. For example, cell overlap may be disallowed [Taylor 84], or devices may not be created by cell overlap [Scheffer 81, Tucker 82, Ousterhout 84a, Ousterhout 85]. In those systems allowing unrestricted cell overlap, arbitrary overlap resulted in severe performance penalties. In the worst case, manipulating arbitrary hierarchies of rectangles is NP-hard, while hierarchies with non-overlapping cells can be manipulated in $n \log n$ time, where n is the number of rectangles [Bentley 81].

9.2.1.3. Using More Space

VLASIC uses about 1750 bytes of storage per transistor or 5.25 megabytes for a 3000 transistor circuit. Assuming 32 megabytes as the practical limit on memory usage, VLASIC is limited to handling circuits with no more than 18,000 transistors. This limit precludes the simulation of most chip designs. A hierarchical description of the 3000 transistor circuit takes 492 K-bytes, or only 164 bytes per transistor, more than a factor of ten smaller than the fully-instantiated circuit. Hierarchy also reduces the size of input files. The fully-instantiated description of the 3000 transistor circuit takes 2.4 megabytes for a text file and 780 K-bytes for a packed binary file. If we assume 32 megabytes as the practical limit for an input file, then VLASIC is limited to designs with no more than 123,000 transistors. This limit is much higher than the limit due to virtual memory usage. A hierarchical description of the 3000 transistor circuit takes only 309 K-bytes for a text file, nearly eight times smaller.

The need to minimize space usage strongly affects the VLASIC layout data structures. For example, the layout bins are 32 microns square when 16 microns square would be more ideal. The bins should be approximately the same size as the average defect. In our simulations, the average defect is never more than 8-10 microns in diameter, so the bins should be much smaller than the current size. Smaller bins reduces the size of the defect neighborhood, resulting in fewer neighboring polygons to examine during fault analysis. The need to look at more polygons in large bins is mitigated by the fact that the polygon operations check bounding boxes before proceeding with additional work.

As was seen in Chapter 6, the open circuit fault analysis procedure is quite expensive. One of the major costs in this procedure is traversing broken nets. Currently the next polygon on the net is determined using intersection tests. A much faster method would be to link all net polygons by pointers. Polygons that touch transistors or bin boundaries could have pointers to additional records. Net traversal would be done by looking at these pointers, with no polygon operations necessary.

The VLASIC polygon package must sacrifice performance to save space. Rectangular polygons are compressed for storage and re-expanded for use. This adds a significant performance penalty. A large part of the fault analysis time goes to merging polygons in the defect neighborhood. As noted in Chapter 6, we do not do the merging in the preprocessing phase because this would result in polygons with many edges. These polygons are large, and slow to process. With more space, it is possible not only to store the many polygon edges, but to store a tree of edge bounding boxes. This tree allows large polygons to be processed in $n \log n$ time for polygons with n edges, so large polygons are not much more expensive to process than small ones. Since the polygon merging need only be done during preprocessing, the fault analysis time is greatly reduced.

An alternative to storing a bounding box tree is to record what polygon edges lie in each bin. For all practical cases, this makes polygon operations constant time, since there are roughly a constant number of polygon edges in the defect vicinity. This edge scheme also has the advantage that only the parts of the result polygon located in the defect neighborhood need be generated, not the entire polygon. We are not interested in the result polygon beyond the neighboring bin boundaries.

9.2.1.4. Fast Simple Fault Analysis

As discussed in Chapter 5, our research originally explored a fault analysis technique based on using the ACE circuit extractor to extract the layout in the defect neighborhood first with, and then without the defect present. The two circuits are then compared to determine what faults have occurred. This technique did not appear to be promising due to the long circuit extraction time (0.5 seconds) and the need to determine the differences between the two extracted circuit graphs. Further examination of this method indicates that major performance improvements are possible by both speeding up the circuit extraction and the comparison phase.

The circuit extraction can be sped up by making the extractor a procedure call instead of a subprocess. The extractor could also return a data structure rather than a text string representing the circuit. The extraction process can also be sped up by using a circuit extractor specialized to extracting only defect neighborhoods [Chew 87].

A major speed improvement can be obtained by using specialized geometric processing. The number of rectangles in the defect approximation can be greatly reduced by using the configuration shown in Figure 9-1.

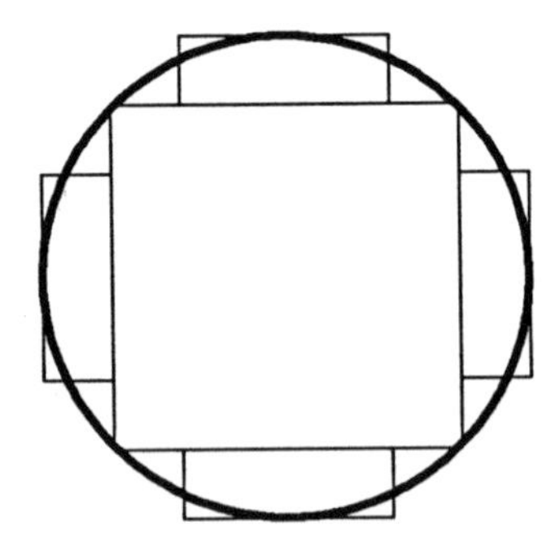

Figure 9-1: Six Rectangle Approximation of a Circle

The use of six rectangles instead of the 21 used by ACE for a ten-micron defect reduces the extraction time by 25%. This approximation has the same horizontal and vertical faces as the octagon, and the same diameter along the diagonal axes. While the octagon area was 5.5% larger than the circle, the six-rectangle approximation has less area. The octagon diameter is 1.0 to 1.08 times that of the circle. The six-rectangle approximation area is 5.4% smaller than the circle area. The diameter is 0.82 to 1.08 times that of the circle.

An alternative technique is to use trapezoids instead of rectangles. The defect octagon can be described by three trapezoids. Trapezoids can be processed almost as quickly as rectangles, and are suitable for representing industrial layout designs [Tarolli 81, Tarolli 83]. Since the defects are small, it is also possible to represent them and the surrounding geometry by a bit map, and perform raster-style circuit extraction [Baker 80].

The combination of these techniques should reduce the extraction time to about 50 msec. Similar performance has already been achieved in an extractor using advanced data structures [Ousterhout 84b, Ousterhout 85].

Once the two circuits have been extracted, they must then be compared with a graph isomorphism algorithm. Since the neighborhood around the defect is small, the circuit graphs contain only a few transistors and nets, and the comparison takes only 10-50 msec [Ebeling 83]. Since defects have some maximum diameter x_m, there is some maximum circuit size in the defect neighborhood, and so a finite number of circuits. For each of the good circuits, there are a finite number of faulty circuits that can occur. In practice there are no more than a few tens of thousands of combinations of good and faulty circuits. These combinations can be stored in a table, along with the fault group represented by the transformation from the good to faulty circuit. The two extracted circuits are used to hash into the table and obtain the fault group. A similar technique has been used for diagnosing global defects [Odryna 85a, Odryna 85b].

By using all of the above techniques, it should be possible to do fault analysis in about 60 msec. This is half the speed of VLASIC, but the simulator would be much simpler.

9.2.2. Non-Monte Carlo Analysis Techniques

A Monte Carlo approach to yield simulation will always be expensive, even if the techniques described in the previous sections are used. We would like to use much cheaper analytic techniques to produce a set of unique fault lists similar to VLASIC output. The basic approach is to estimate the critical areas for each circuit fault group as a function of the defect type and size. Recall that a fault group is those faults that occur due to one defect. We can calculate the probability of each group occurring when a defect drawn from its size distribution lands in the cell. We then use the defect spatial distribution to calculate the probability of each unique fault list occurring for a collection of cells forming a chip. We first describe the critical area analysis, including previous work. We then describe how to calculate the joint fault probabilities for a cell, and then how to combine the cells with the defect spatial distribution to obtain the probability of each unique fault list occurring.

9.2.2.1. Critical Area Analysis

Since oxide pinholes and junction leakage are modeled as points, the critical areas for them are the dielectric between overlapping conductors, and the junction areas, respectively [Stapper 80]. For extra and missing material defects, the critical area is a function of the defect size distribution [Stapper 76]. Most researchers have assumed a size distribution in order to calculate an effective critical area. We would like to determine a critical area function so that different size distributions can be substituted at a later time.

Stapper et al. estimated the critical areas for a redundant DRAM using a sampling procedure [Stapper 80]. Random coordinates were marked on 2000X plots. Circular extra and missing material defects of different sizes were placed on these coordinates and visually examined to determine what type of circuit failure (i.e. failed row, bit, column), if any, had occurred, producing an estimate of the probability of failure versus defect size for each defect type. This result was combined with a $1/x^3$ size distribution to determine the overall probability of failure.

VLASIC can be used to automate this sampling process. In Section 7.3.1 VLASIC was used to determine the yield of a single DRAM cell, and this result was combined with the spatial statistics to determine the yield of a DRAM array. The results were poor because they neglected cell boundary effects. The results could be improved by using clusters of cells, thus taking into account the effects of neighboring cells. VLASIC can also be used to calculate the critical area by simply multiplying the fraction of defects that cause circuit faults by the cell area.

One advantage of critical area analysis is that importance sampling can be used to reduce the number of defects placed and analyzed. Those cells that make up a large fraction of the chip area can be heavily sampled to obtain an accurate critical area estimation, while small cells that are only used once or twice can be lightly sampled, since an error in the critical area does not significantly affect the yield of the chip.

In addition to sampling on the basis of cell frequency, it is also possible to do importance sampling based on the probability of failure (POF). Small extra and missing material defects do not cause circuit faults, as discussed in Section 4.1. Similarly, very large defects always cause a circuit fault. The region of interest is between these two extremes. The effective size distribution can be distorted by multiplying it by the POF density function to minimize the number of samples required for a given level of accuracy. The POF density function is not known prior to the start of simulation. It can be initially approximated as a rectangle or triangle, as was shown in Figure 2-1 and then refined as the simulation proceeds.

The truncated size distribution shown in Figure 4-2 is a simple example of a distorted size distribution. As was discussed in Section 6.10.1, the truncated size distribution gains little in performance since fault analysis for small defects is very cheap. Similarly, if one is interested in determining only that a circuit fault has occurred, but not the details, then analysis of large defects is also low in cost. A distorted size distribution can reduce sampling time significantly only if it substantially reduces the sample size for defects in the size range between very large and very small defects.

Stapper noted that since each defect type was independent, separate critical areas had to be calculated for each type [Stapper 83b]. He showed that the critical area for shorting two lines or breaking one line increases linearly with defect diameter, taking into account the fact that lines must have minimum width Δw and spacing Δs to avoid opens and shorts. This work was later extended to multiple parallel lines, where adjacent critical areas can expand until they merge together [Stapper 84a]. Others have developed a similar multi-line model [Ferris-Prabhu 85a, Ferris-Prabhu 85b, Ferris-Prabhu 85c, Maly 86a].

Maly and Deszczka developed a single-line model for arbitrary shapes, using polygon operations to calculate critical areas [Maly 83]. The critical area for short circuits was calculated by expanding the polygons by half the defect diameter and intersecting them as shown in Figure 9-2.

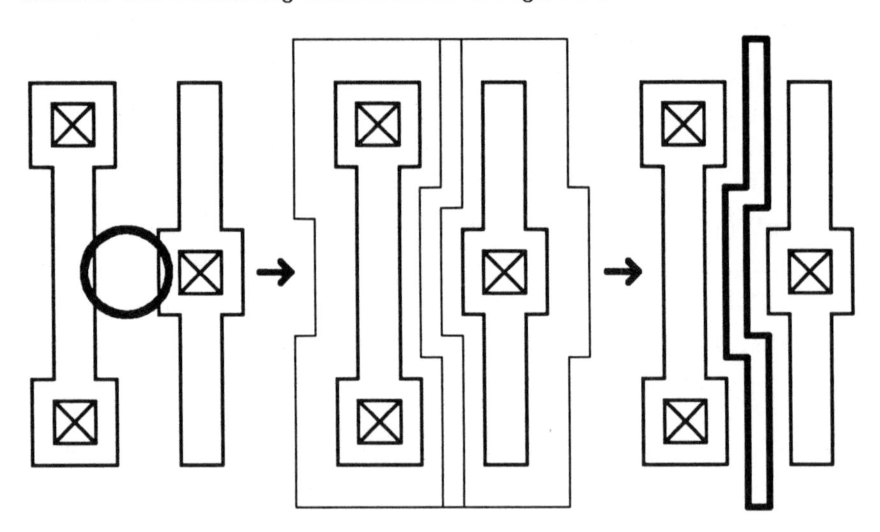

Figure 9-2: Critical Area for Short

Open circuits could be considered breaks in the space between conductors, and so the critical area was calculated by expanding the space and self-

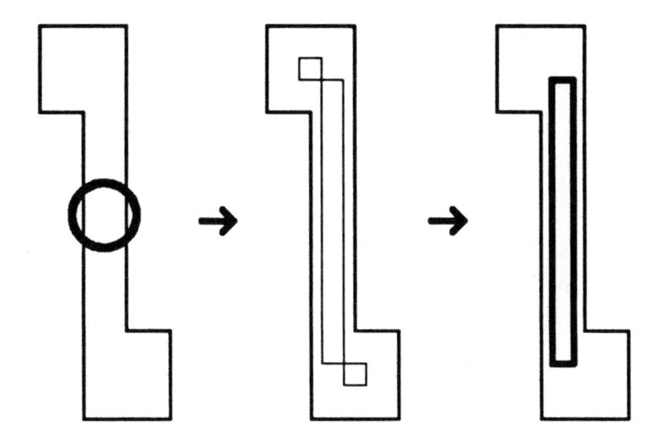

Figure 9-3: Critical Area for Open

intersecting as shown in Figure 9-3. We have described a similar technique [Walker 82, Walker 83].

Maly used a multi-line model for estimating critical area [Maly 84b, Maly 85b, Maly 85c, Maly 85a]. Rather than modeling the chip as a set of identical parallel lines, a *virtual artwork* was used. The virtual artwork is a set of lines that have the same statistical characteristics as the actual layout, in terms of total length of lines, average spacing, area of active elements, length of lines at a given width and spacing, etc. The statistics were used to generate an artwork in a form convenient for analysis, such as a set of parallel lines with the combinations of lengths, widths, and spacings matching the statistics. A system to extract the statistical characteristics of a layout is now commercially available [Perry 85].

Neighboring cells can have critical areas associated with their interactions, for example, a critical area for shorting between lines in two cells. The literature has not discussed this point. When determining the critical area for a chip, it is necessary to calculate the critical area for all cell interactions as well as all individual cells. The set of unique areas can be determined by examining the cell hierarchy [Hon 80, Hon 83].

9.2.2.2. Fault Probabilities and Fault Lists

Once we have obtained the critical areas, we can calculate the fault group probabilities for a defect landing in the cell based on the size distribution. The areas as a function of size are convolved with the size distribution for each defect type in order to obtain the effective critical area for each fault group. Assuming that defects within a cell have a uniform spatial distribution, the probability of each fault group is simply the fraction of the cell area taken up by the effective critical area for that group. This analysis makes the assumption that defects do not interfere with one another, as discussed in Chapter 3.

We assume that defects are placed uniformly within the die. Therefore the probability that a defect lands on any cell is just the fraction of chip area taken up by that cell. The probability of a defect landing on a cell is multiplied by the probability of each fault group within the cell to obtain the probability of each fault group occurring if a defect lands on the chip. The defect spatial distribution is used to determine the probability of each defect count occurring on the chip, and from that the probability of each combination of fault groups.

The number of unique fault groups grows exponentially with circuit size, so it is not possible to calculate the probabilities for all of them in practical cases. For most applications, it is adequate to place fault groups into equivalence classes. For example, in the case of a memory, all fault groups causing a single-bit failure can be treated the same.

The critical area analysis technique has the advantage that it uses the cell hierarchy. The critical areas for each cell and cell overlap region need only be calculated once. If cells are replicated in the design, or used in several designs, no additional work need be done. A second advantage is that fault groups can be merged into equivalence classes before unique fault lists are generated. A third advantage is that the probability of rare faults can be determined much more readily than with a Monte Carlo technique, which requires a very large number of trials to obtain a reasonable confidence interval. Finally, the critical areas can be made a function of global defects, as well as local defects, allowing both classes of defects to be combined, as discussed below.

9.2.3. Combining Local and Global Defects

As discussed in Chapter 1, local and global defects and the circuit faults caused by them are not independent. Misalignment modifies the overlap area between conductors on different layers, which in turn modifies the probability of an oxide pinhole defect causing a short circuit. An increase in line width increases the chances that an extra material defect will cause a short circuit, and reduces the chances that a missing material defect will cause an open circuit. As discussed in Chapter 3, some global defects, such as poor step coverage, result in DC changes to the circuit topology. At the same time, local defects affect parametric behavior. Extra or missing material can modify device sizes, and change capacitances and resistances. We would like to combine both local and global defects in one yield simulator [Maly 86b, Chen 86, Chew 87, Chen 87].

An important feature of the FABRICS II statistical process and device simulator is that the fabrication process is described as a sequence of process steps, and global defects are described in terms of disturbances to each of these steps. Analytic models are used to convert disturbed process conditions into device parameters. We have described local defects directly in terms of modifications to the layout geometry. We would like to be able to describe the disturbances that cause local defects, such as photoresist pinholes, dust particles, and crystal defects, and use models to translate these into layout modifications. Some work has already been done in this area by convolving a defect size distribution with the line width variation introduced in the lithography process [Maly 85c]. One difficulty with this approach is that the physics of how process disturbances cause local defects (e.g. a substrate imperfection causing a gate oxide pinhole) is much less well understood than for global defects.

References

[Abbott 81] R. Abbott, K. Kokkonen, R. Kung, and R. Smith.
Equipping a Line of Memories with Spare Cells.
Electronics 54(15):127-130, July 28, 1981.

[Ahrens 74] J. Ahrens and U. Dieter.
Computer Methods for Sampling from Gamma, Beta, Poisson and Binomial Distributions.
Computing 12(3):223-246, 1974.

[Ahrens 80] J. Ahrens and U. Dieter.
Sampling from Binomial and Poisson Distributions: A Method with Bounded Computation Times.
Computing 25(3):193-208, 1980.

[Ansley 68] W. Ansley.
Computation of Integrated-Circuit Yields from the Distribution of Slice Yields for the Individual Devices.
IEEE Transactions on Electron Devices ED-15(6):405-406, June, 1968.

[Baker 80] C. Baker and C. Terman.
Tools for Verifying Integrated Circuit Designs.
Lambda (now VLSI Design) 1(3):22-31, Fourth Quarter, 1980.

[Barton 79] E. Barton.
The Polygon Package.
Silicon Structures Project 3129, California Institute of Technology, Computer Science Department, November, 1979.

[Barton 80a] A. Barton.
A Fault Tolerant Integrated Circuit Memory.
PhD thesis, California Institute of Technology, Computer Science Department, April, 1980.

[Barton 80b] E. Barton and I. Buchanan.
The Polygon Package.
Computer-Aided Design 12(1):3-11, January, 1980.

[Benevit 82] C. Benevit, J. Cassard, K. Dimmler, A. Dumbri, M. Mound, F. Procyk, W. Rosenzweig, and A. Yanof.
256K Dynamic Random Access Memory.
In *IEEE International Solid-State Circuits Conference Digest of Technical Papers*, pages 76-77. February, 1982.

[Bentley 80] J. Bentley, D. Haken, and R. Hon.
Statistics on VLSI Designs.
Technical Report CMU-CS-80-111, Carnegie Mellon University, Computer Science Department, April, 1980.

[Bentley 81] J. Bentley and T. Ottmann.
The Complexity of Manipulating Hierarchically Defined Sets of Rectangles.
Technical Report CMU-CS-81-109, Carnegie Mellon University, Computer Science Department, April, 1981.

[Bernard 78] J. Bernard.
The IC Yield Problem: A Tentative Analysis for MOS/SOS Circuits.
IEEE Transactions on Electron Devices ED-25(8):939-944, August, 1978.

[Bertram 83] W. Bertram.
Yield and Reliability.
VLSI Technology.
McGraw-Hill, New York, 1983, pages 599-612, Chapter 14.

[Borisov 78] V. Borisov, V. Konopel'ko, and V. Losev.
Use of Redundancy for Increasing the Reliability and Percentage Yield of Memory-Related Circuits.
Mikroelektronika 7(4):250-257, July-August, 1978.

[Borisov 79] V. Borisov.
A Probability Method for Estimating The Effectiveness of Redundancy in Semiconductor Memory Structures.
Mikroelektronika 8(3):280-282, May-June, 1979.

[Buehler 81] M. Buehler and L. Linholm.
Role of Test Chips in Coordinating Logic and Circuit Design and Layout Aids for VLSI.
Solid State Technology 24(9):68-73, September, 1981.

[Buehler 83] M. Buehler, B. Blaes, C. Pina, and T. Griswold.
Pinhole Array Capacitor for Oxide Integrity Analysis.
Solid State Technology 26(11):131-137, November, 1983.

[Carver 80] G. Carver, L. Linholm, and T. Russell.
Use of Microelectronic Test Structures to Characterize IC Materials, Processes, and Processing Equipment.
Solid State Technology 23(9):85-92, September, 1980.

[Cenker 79] R. Cenker, D. Clemons, W. Huber, J. Petrizzi, F. Procyk, and G. Trout.
A Fault-Tolerant 64K Dynamic Random-Access Memory.
IEEE Transactions on Electron Devices ED-26(6):853-860, June, 1979.

[Chen 69] A. Chen.
Redundancy in LSI Memory Array.
IEEE Journal of Solid-State Circuits SC-4(5):291-293, October, 1969.

[Chen 86] I. Chen and A. J. Strojwas.
Realistic Yield Simulation for IC Structural Failures.
In *IEEE International Conference on Computer-Aided Design (ICCAD) Digest of Technical Papers*, pages 220-223. IEEE, November, 1986.

[Chen 87] I. Chen.
Realistic Yield Simulation for VLSI Structural Failures.
Thesis Proposal, Carnegie Mellon University, Electrical and Computer Engineering Department, April, 1987.

[Chew 87] M. Chew.
A Methodology for the Efficient Circuit Extraction of Process Deformed Layouts.
Master's thesis, Carnegie Mellon University, Electrical and Computer Engineering Department, February, 1987.

[Chwang 83] R. Chwang, M. Choi, D. Creek, S. Stern, P. Pelley, J. Schutz, P. Warkentin, M. Bohr, and K. Yu.
A 70ns High Density 64K CMOS Dynamic RAM.
IEEE Journal of Solid-State Circuits SC-18(5):457-463, October, 1983.

[Cleverley 83] D. Cleverley.
Product Quality Level Monitoring and Control for Logic Chips and Modules.
IBM Journal of Research and Development 27(1):4-10, January, 1983.

[Davis 85] H. Davis.
A 70-ns Word-Wide 1-Mbit ROM With On-Chip Error-Correction Circuits.
IEEE Journal of Solid-State Circuits SC-20(5):958-963, October, 1985.

[de la Rosa 86] J. de la Rosa and D. Rose.
Wafer Inspection with a Laser Scanning Microscope.
AT&T Technical Journal 65(1):68-77, January-February, 1986.

[DeGroot 75] M. DeGroot.
Probability and Statistics.
Addison-Wesley, Reading, MA, 1975.

[Dingwall 68] A. Dingwall.
High-Yield Processing for Fixed-Interconnect Large-Scale Integrated Arrays.
IEEE Transactions on Electron Devices ED-15(9):631-637, September, 1968.

[Ebel 82] A. Ebel, G. Atwood, E. So, S. Liu, N. Kynett, R. Jecmen, J. Mingo, and H. Dun.
A NMOS 64K Static RAM.
In *IEEE International Solid-State Circuits Conference Digest of Technical Papers*, pages 254-255,331. February, 1982.

[Ebeling 83] C. Ebeling and O. Zajicek.
Validating VLSI Circuit Layout by Wirelist Comparison.
In *IEEE International Conference on Computer-Aided Design Digest of Technical Papers*, pages 172-173. IEEE, September, 1983.

[Egawa 80] Y. Egawa, T. Wada, Y. Ohmori, N. Tsuda, and K. Masuda.
A 1-Mbit Full-Wafer MOS RAM.
IEEE Journal of Solid-State Circuits SC-15(4):677-686, August, 1980.

[Faust 80a] M. Faust.
Using Polygon Operations to Specify Design Rule Checks.
VLSI Document V048, Carnegie Mellon University, Computer Science Department, May, 1980.

[Faust 80b] M. Faust.
A Polygon Package.
VLSI Document V049, Carnegie Mellon University, Computer Science Department, October, 1980.

[Faust 81a] M. Faust.
Comments on "The Polygon Package".
VLSI Document V072, Carnegie Mellon University, Computer Science Department, January, 1981.

[Faust 81b] M. Faust.
Oracle: A Design Rule Checking Program.
VLSI Document V096, Carnegie Mellon University, Computer Science Department, August, 1981.

[Faust 81c] M. Faust.
Geometric Techniques in VLSI Design and Verification.
VLSI Document V099, Carnegie Mellon University, Computer Science Department, September, 1981.

[Ferguson 87] F. J. Ferguson.
Ph.D. thesis in preparation.
April, 1987.

[Ferris-Prabhu 85a]
A. Ferris-Prabhu.
Modeling the Critical Area in Yield Forecasts.
IEEE Journal of Solid-State Circuits SC-20(4):874-878, August, 1985.

[Ferris-Prabhu 85b]
A. Ferris-Prabhu.
Defect Size Variations and Their Effect on the Critical Area of VLSI Devices.
IEEE Journal of Solid-State Circuits SC-20(4):878-880, August, 1985.

[Ferris-Prabhu 85c]
A. Ferris-Prabhu.
Role of Defect Size Distribution in Yield Modeling.
IEEE Transactions on Electron Devices ED-32(9):1727-1736, September, 1985.

[Ferris-Prabhu 87]
A. Ferris-Prabhu, L. Smith, H. Bonges, and J. Paulsen.
Radial Yield Variations in Semiconductor Wafers.
IEEE Circuits and Devices Magazine 3(2):42-47, March, 1987.

[Fisher 83] A. Fisher, H. Kung, L. Monier, H. Walker, and Y. Dohi.
Design of the PSC: A Programmable Systolic Chip.
In R. Bryant (editor), *Proceedings of the Third Caltech Conference on VLSI*, pages 287-302. Computer Science Press, Rockville, MD, March, 1983.

[Fitzgerald 80] B. Fitzgerald and E. Thoma.
Circuit Implementation of Fusible Redundant Addresses on RAMs for Productivity Enhancement.
IBM Journal of Research and Development 24(3):291-298, May, 1980.

[Fitzgerald 82] B. Fitzgerald and E. Thoma.
A 288Kb Dynamic RAM.
In *IEEE International Solid-State Circuits Conference Digest of Technical Papers*, pages 68-69. February, 1982.

[Frank 81] E. Frank, C. Ebeling, and B. Sproull.
Hierarchical Wirelist Format.
VLSI Document V085, Carnegie Mellon University, Computer Science Department, July, 1981.

[Frank 82] E. Frank.
The Fast-1: A Data-Driven Multiprocessor for Logic Simulation - A Thesis Proposal.
VLSI Document V122, Carnegie Mellon University, Computer Science Department, October, 1982.

[Frank 85] E. Frank.
A Data-Driven Multiprocessor for Switch-Level Simulation of VLSI Circuits.
PhD Thesis CMU-CS-85-180, Carnegie Mellon University, Computer Science Department, November, 1985.

[Fujii 83] S. Fujii, K. Natori, T. Furuyama, S. Saito, H. Toda, T. Tanaka, and O. Ozawa.
A Low-Power Sub 100ns 256K Bit Dynamic RAM.
IEEE Journal of Solid-State Circuits SC-18(5):441-446, October, 1983.

[Gangatirkar 82] P. Gangatirkar, R. Presson, and L. Rosner.
Test/Characterization Procedures for High Density Silicon RAMs.
In *IEEE International Solid-State Circuits Conference Digest of Technical Papers*, pages 62-63. February, 1982.

[Glaser 77] A. Glaser and G. Subak-Sharpe.
Integrated Circuit Engineering.
Addison-Wesley, Reading, MA, 1977.

[Gongwer 83] G. Gongwer and K. Gudger.
A 16K E^2PROM Using E^2 Element Redundancy.
IEEE Journal of Solid-State Circuits SC-18(5):550-553, October, 1983.

[Guibas 83] L. Guibas, L. Ramshaw, and J. Stolfi.
A Kinetic Framework for Computational Geometry.
In *IEEE Foundations of Computer Science Proceedings*, pages 100-111. 1983.

[Gupta 70] A. Gupta and J. Lathrop.
Comments on "Influence of Epitaxial Mounds on the Yield of Integrated Circuits".
Proceedings of the IEEE 58(12):1960-1961, December, 1970.

[Gupta 72] A. Gupta and J. Lathrop.
Yield Analysis of Large Integrated-Circuit Chips.
IEEE Journal of Solid-State Circuits SC-7(5):389-395, October, 1972.

[Gupta 74] A. Gupta, W. Porter, and J. Lathrop.
Defect Analysis and Yield Degradation of Integrated Circuits.
IEEE Journal of Solid-State Circuits SC-9(3):96-103, June, 1974.

[Gupta 81] A. Gupta.
ACE - A Circuit Extractor.
VLSI Document V105, Carnegie Mellon University, Computer Science Department, June, 1981.

[Gupta 83a] A. Gupta.
ACE: A Circuit Extractor.
In *ACM IEEE 20th Design Automation Conference Proceedings.* June, 1983.

[Gupta 83b] A. Gupta and R. Hon.
HEXT: A Hierarchical Circuit Extractor.
Journal of VLSI and Computer Systems 1(1):23-39, 1983.

[Gutierrez 84] J.-M. Gutierrez.
Private Communication.
June, 1984.

[Ham 78] W. Ham.
Yield-Area Analysis: Part I - A Diagnostic Tool for Fundamental Integrated-Circuit Process Problems.
RCA Review 39(2):231-249, June, 1978.

[Haraszti 82] T. Haraszti.
A Novel Associative Approach for Fault-Tolerant MOS RAM's.
IEEE Journal of Solid-State Circuits SC-17(3):539-546, June, 1982.

[Hardee 81] K. Hardee and R. Sud.
A Fault-Tolerant 30 ns/375 mW 16K x 1 NMOS Static RAM.
IEEE Journal of Solid-State Circuits SC-17(5):435-443, October, 1981.

[Harden 86] J. Harden and T. Mangir.
Comments on "Sources of Failures and Yield Improvement for VLSI and Restructurable Interconnects for RVLSI and WSI: Part I".
Proceedings of the IEEE 74(3):515-516, March, 1986.

[Hemmert 81] R. Hemmert.
Poisson Process and Integrated Circuit Yield Prediction.
Solid-State Electronics 24(6):511-515, June, 1981.

[Hofstein 63] S. Hofstein and F. Heiman.
The Silicon Insulated-Gate Field-Effect Transistor.
Proceedings of the IEEE 51(9):1190-1202, September, 1963.

[Hon 80] B. Hon.
The Hierarchical Analysis of VLSI Designs.
VLSI Document V073, Carnegie Mellon University, Computer Science Department, December, 1980.

[Hon 83] R. Hon.
The Hierarchical Analysis of VLSI Designs.
PhD thesis, Carnegie Mellon University, Computer Science Department, December, 1983.

[Hu 79] S. Hu.
Some Considerations in the Formulation of IC Yield Statistics.
Solid-State Electronics 22(2):205-211, February, 1979.

[Ipri 77] A. Ipri and J. Sarace.
Integrated Circuit Process and Design Rule Evaluation Techniques.
RCA Review 38(3):323-350, September, 1977.

[Ipri 79] A. Ipri.
Impact of Design Rule Reduction on Size, Yield, and Cost of Integrated Circuits.
Solid State Technology 22(2):85-91, February, 1979.

[Ipri 80] A. Ipri.
Evaluation of CMOS Transistor Related Design Rules.
RCA Review 41(4):537-549, December, 1980.

[Ishihara 82] M. Ishihara, T. Matsumoto, S. Shimizu, K. Mitsusada, and K. Shimohigashi.
A 256K Dynamic MOS RAM with Alpha Immune and Redundancy.
In *IEEE International Solid-State Circuits Conference Digest of Technical Papers*, pages 74-75. February, 1982.

[Jastrzebski 82] L. Jastrzebski.
Origin and Control of Material Defects in Silicon VLSI Technologies: An Overview.
IEEE Journal of Solid-State Circuits SC-17(2):105-117, April, 1982.

[Jerdonek 78] R. Jerdonek, H. Bare, and G. Fromen.
The Use of a Silicon-Gate C-MOS/SOS Test Vehicle to Evaluate Technology Maturity.
IEEE Transactions on Electron Devices ED-25(8):873-878, August, 1978.

[Johnson 69] N. Johnson and S. Kotz.
Discrete Distributions.
Houghton Mifflin, Boston, 1969.

[Jouppi 83] N. Jouppi.
TV: An nMOS Timing Analyzer.
In *Proceedings of the Third Caltech Conference on VLSI*, pages 71-85. Computer Science Press, March, 1983.

[Ketchen 85] M. Ketchen.
Point Defect Yield Model for Wafer Scale Integration.
IEEE Circuits and Devices Magazine 1(4):24-34, July, 1985.

[Khurana 82] N. Khurana.
Redundancy - A Technique for Cost Effective VLSI.
1982.
Slides of a talk at Carnegie Mellon University.

[Kim 78] C. Kim and W. Ham.
Yield-Area Analysis: Part II - Effects of Photomask Alignment Errors on Zero Yield Loci.
RCA Review 39(4):565-576, December, 1978.

[Kitano 80] Y. Kitano, S. Kohda, H. Kikuchi, and S. Sakai.
A 4-Mbit Full-Wafer ROM.
IEEE Journal of Solid-State Circuits SC-15(4):687-693, August, 1980.

[Knuth 81] D. Knuth.
The Art of Computer Programming. Volume 2: *Seminumerical Algorithms.*
Addison-Wesley, Reading, MA, 1981.

[Koren 84] I. Koren and M. Breuer.
On Area and Yield Considerations for Fault-Tolerant VLSI Processor Arrays.
IEEE Transactions on Computers C-33(1):21-27, January, 1984.

[Kovchavtsev 79] A. Kovchavtsev and A. Frantsuzov.
Defect Density in Thermally Grown Silicon Dioxide with Thicknesses 30-600 Angstroms.
Mikroelektronika 8(5):330-333, September-October, 1979.

[Kung 84] H. Kung and M. Lam.
Wafer-Scale Integration and Two-Level Pipelined Implementations of Systolic Arrays.
Journal of Parallel and Distributed Computing 1(1):32-63, 1984.

[Lang 79] D. Lang.
Private Communication.
1979.

[Lashevskii 79] R. Lashevskii, G. Filaretov, L. Chasovnikova, and L. Shapiro.
Prediction of the Yield of Suitable MOS Structures with Respect to the Subgate Dielectric.
Mikroelektronika 8(4):281-283, July-August, 1979.

[Lawson 66] T. Lawson.
A Prediction of the Photoresist Influence on Integrated Circuit Yield.
Solid State Technology 9(7):22-25, July, 1966.

[Lee 84] D. Lee and F. Preparata.
Computational Geometry - A Survey.
IEEE Transactions on Computers C-33(12):1072-1101, December, 1984.

[Linholm 81] L. Linholm.
Semiconductor Measurement Technology: The Design, Testing, and Analysis of a Comprehensive Test Pattern for Measuring CMOS/SOS Process Performance and Control.
Special Publication, National Bureau of Standards, 1981.

[Losleben 79] P. Losleben and K. Thompson.
Topological Analysis for VLSI Circuits.
In *ACM IEEE 16th Design Automation Conference Proceedings*, pages 461-473. June, 1979.

[Lukacs 72] E. Lukacs.
Probability and Mathematical Statistics.
Academic Press, New York, 1972.

[Mallory 83] C. Mallory, D. Perloff, T. Hasan, and R. Stanley.
Spatial Yield Analysis in Integrated Circuit Manufacturing.
Solid State Technology 26(11):121-127, November, 1983.

[Maly 81] W. Maly, A. J. Strojwas, and S. W. Director.
Fabrication Based Statistical Design of Monolithic ICs.
In *Proceedings of the IEEE International Symposium on Circuits and Systems*, pages 135-138. April, 1981.

[Maly 82a] W. Maly and A. J. Strojwas.
Statistical Simulation of the IC Manufacturing Process.
IEEE Transactions on Computer-Aided Design of Integrated Circuits and Systems 1(3), July, 1982.

[Maly 82b] W. Maly and M. Marek-Sadowska.
Artwork Analysis using Hierarchical Layout Description.
1982.
Submitted to ICCC82.

[Maly 83] W. Maly and J. Deszczka.
Yield Estimation Model for VLSI Artwork Evaluation.
Electronics Letters 19(6):226-227, March 17, 1983.

[Maly 84a] W. Maly, F. J. Ferguson, and J. P. Shen.
Systematic Characterization of Physical Defects for Fault Analysis of MOS IC Cells.
In *IEEE International Test Conference Proceedings*, pages 390-399. October, 1984.

[Maly 84b] W. Maly.
Modeling of Point Defect Related Yield Losses for CAD of VLSI Circuits.
In *IEEE International Conference on Computer-Aided Design Digest of Technical Papers*, pages 161-163. November, 1984.

[Maly 85a] W. Maly.
Private Communication.
September, 1985.

[Maly 85b] W. Maly.
Topics in Point Defect Related Yield Losses.
Technical Report CMUCAD-85-41, Carnegie Mellon University, Electrical and Computer Engineering Department, February, 1985.

[Maly 85c] W. Maly.
Modeling of Lithography Related Yield Losses for CAD of VLSI Circuits.
IEEE Transactions on Computer-Aided Design of Integrated Circuits and Systems CAD-4(3):166-177, July, 1985.

[Maly 86a] W. Maly.
Design Automation Seminar Talk.
April, 1986.

[Maly 86b] W. Maly, A. J. Strojwas and S. W. Director.
VLSI Yield Prediction and Estimation: A Unified Framework.
IEEE Transactions on Computer-Aided Design of Integrated Circuits and Systems CAD-5(1):114-130, January, 1986.

[Maly 87a] W. Maly.
Realistic Fault Modeling for VLSI Testing.
In *ACM IEEE 24th Design Automation Conference Proceedings.* June, 1987.

[Maly 87b] Wojciech Maly, Michael E. Thomas, and Jeffrey D. Chinn.
Double-Bridge Test Structure for the Evaluation of Type, Size and Density of Spot Defects.
Technical Report CMUCAD-87-2, CMU, February, 1987.

[Mangir 82] T. Mangir and A. Avizienis.
Fault-Tolerant Design for VLSI: Effect of Interconnect Requirements on Yield Improvement of VLSI Designs.
IEEE Transactions on Computers C-31(7):609-615, July, 1982.

[Mangir 84] T. Mangir.
Sources of Failures and Yield Improvement for VLSI and Restructurable Interconnects for RVLSI and WSI: Part I - Sources of Failures and Yield Improvement for VLSI.
Proceedings of the IEEE 72(6):690-708, June, 1984.

[Mano 80] T. Mano, K. Takeya, W. Takashi, N. Ieda, K. Kiuchi, E. Arai, T. Ogawa, and K. Hirata.
A Fault-Tolerant 256K RAM Fabricated with Molybdenum-Polysilicon Technology.
IEEE Journal of Solid-State Circuits SC-15(5):865-872, October, 1980.

[Mano 82] T. Mano, M. Wada, N. Ieda, and M. Tanimoto.
A Redundancy Circuit for a Fault-Tolerant 256K MOS RAM.
IEEE Journal of Solid-State Circuits SC-17(4):726-731, August, 1982.

[Marek-Sadowska 82]
M. Marek-Sadowska and W. Maly.
A Hierarchical Layout Description for Artwork Analysis of VLSI IC.
In *IEEE International Conference on Circuits and Computers Proceedings,* pages 419-422. September, 1982.

[McCanny 83] J. McCanny and J. McWhirter.
Yield Enhancement of Bit Level Systolic Array Chips Using Fault Tolerant Techniques.
Electronics Letters 19(14):525-527, July 7, 1983.

[McGrath 80] E. McGrath and T. Whitney.
Design Integrity and Immunity Checking: A New Look at Layout Verification and Design Rule Checking.
In *ACM IEEE 17th Design Automation Conference Proceedings*, pages 263-268. June, 1980.

[McMinn 82] S. McMinn.
Semiconductor Manufacturing Considerations for VLSI Designers.
VLSI Design 3(4):16-18, July/August, 1982.

[Meister 83] B. Meister.
On Murphy's Yield Formula.
IBM Journal of Research and Development 27(6):545-548, November, 1983.

[Milenkovic 85] V. Milenkovic.
Private Communication.
October, 1985.

[Minato 82] O. Minato, T. Masuhara, T. Sasaki, Y. Sakai, T. Hayashida, K. Nagasawa, K. Nishimura, and T. Yasui.
A Hi-CMOSII 8K x 8 Bit Static RAM.
IEEE Journal of Solid-State Circuits SC-17(5):793-798, October, 1982.

[Mitchell 85] M. Mitchell.
Defect Test Structures for Characterization of VLSI Technologies.
Solid State Technology 28(5):207-213, May, 1985.

[Moore 70] G. Moore.
What Level of LSI is Best for You?
Electronics 43(4):126-130, February 16, 1970.

[Moore 84a] W. Moore and M. Day.
Yield-Enhancement of a Large Systolic Array Chip.
Microelectronics and Reliability 24(3):511-526, 1984.

[Moore 84b] W. Moore.
Switching Circuits for Yield-Enhancement of an Array Chip.
Electronics Letters 20(16):667-669, August 2, 1984.

[Muehldorf 75] E. Muehldorf.
Fault Clustering: Modeling and Observation on Experimental LSI Chips.
IEEE Journal of Solid-State Circuits SC-10(4):237-244, August, 1975.

[Murphy 64] B. Murphy.
Cost-Size Optima of Monolithic Integrated Circuits.
Proceedings of the IEEE 52(12):1537-1545, December, 1964.

[Murphy 71] B. Murphy.
Comments on "A New Look at Yield of Integrated Circuits".
Proceedings of the IEEE 59(8):1128, July, 1971.

[Nassif 82] S. R. Nassif, A. J. Strojwas, and S. W. Director.
FABRICS II: A Statistical Simulator of the IC Fabrication Process.
In *IEEE International Conference on Circuits and Computers Proceedings*, pages 298-301. September, 1982.

[Nassif 84a] S. R. Nassif, A. J. Strojwas and S. W. Director.
FABRICS II: A Statistically Based IC Fabrication Process Simulator.
IEEE Transactions on Computer-Aided Design of Integrated Circuits and Systems CAD-3(1):40-46, January, 1984.

[Nassif 84b] S. R. Nassif.
Private Communication.
1984.

[Newell 82] M. Newell and D. Fitzpatrick.
Exploiting Structure in Integrated Circuit Design Analysis.
In *Proceedings of the 1982 MIT Conference on Advanced Research in VLSI*, pages 84-92. January, 1982.

[Noice 81] D. Noice, J. Newkirk, and R. Mathews.
A Polygon Package for Analyzing Integrated Circuit Designs.
VLSI Design 2(3):33-36, Third Quarter, 1981.

[Ochii 82] K. Ochii, K. Hashimoto, H. Yasuda, M. Masuda, T. Kondo, H. Nozawa, and S. Kohyama.
An Ultralow Power 8K x 8-Bit Full CMOS RAM with a Six-Transistor Cell.
IEEE Journal of Solid-State Circuits SC-17(5):798-803, October, 1982.

[Odryna 85a] P. Odryna.
A VLSI Fault Diagnosis System.
Master's thesis, Carnegie Mellon University, Electrical and Computer Engineering Department, May, 1985.

[Odryna 85b] P. Odryna and A. J. Strojwas.
PROD: A VLSI Fault Diagnosis System.
IEEE Design and Test of Computers 2(6):27-35, December, 1985.

[Okabe 72] T. Okabe, M. Nagata and S. Shimada.
Analysis on Yield of Integrated Circuits and a New Expression for the Yield.
Electrical Engineering in Japan 92:135-141, December, 1972.

[Ousterhout 83] J. Ousterhout.
Crystal: A Timing Analyzer for nMOS VLSI Circuits.
In *Proceedings of the Third Caltech Conference on VLSI*, pages 57-69. Computer Science Press, March, 1983.

[Ousterhout 84a] J. Ousterhout, G. Hamachi, R. Mayo, W. Scott, and G. Taylor.
Magic: A VLSI Layout System.
In *ACM IEEE 21st Design Automation Conference Proceedings*, pages 152-159. June, 1984.

[Ousterhout 84b] J. Ousterhout.
Corner Stitching: A Data-Structuring Technique for VLSI Layout Tools.
IEEE Transactions on Computer-Aided Design of Integrated Circuits and Systems. CAD-3(1):87-100, January, 1984.

[Ousterhout 85] J. Ousterhout, G. Hamachi, R. Mayo, W. Scott, and G. Taylor.
The Magic VLSI Layout System.
IEEE Design and Test of Computers 2(1):19-30, February, 1985.

[Paz 77] O. Paz and T. Lawson.
Modification of Poisson Statistics: Modeling Defects Induced by Diffusion.
IEEE Journal of Solid-State Circuits SC-12(5):540-546, October, 1977.

[Peltzer 83] D. Peltzer.
Wafer-Scale Integration: The Limits of VLSI?
VLSI Design 4(5):43-47, September, 1983.

[Perry 85] S. Perry, M. Mitchell, and D. Pilling.
Yield Analysis Modeling.
In *ACM IEEE 22nd Design Automation Conference Proceedings*, pages 425-428. June, 1985.

[Petritz 67] R. Petritz.
Current Status of Large Scale Integration Technology.
IEEE Journal of Solid-State Circuits SC-2(4):130-147, December, 1967.

[Posa 81] J. Posa.
What to Do When the Bits Go Out.
Electronics 54(15):117-120, July 28, 1981.

[Price 70] J. Price.
A New Look at Yield of Integrated Circuits.
Proceedings of the IEEE 58(8):1290-1291, August, 1970.

[Raffel 85] J. Raffel, A. Anderson, G. Chapman, K. Konkle, B. Mathur, A. Soares, and P. Wyatt.
A Wafer-Scale Digital Integrator Using Restructurable VLSI.
IEEE Journal of Solid-State Circuits SC-20(1):399-406, February, 1985.

[Razdan 85] R. Razdan and A. J. Strojwas.
Statistical Design Rule Developer.
In *IEEE International Conference on Computer-Aided Design*, pages 315-317. November, 1985.

[Rung 81] R. Rung.
Determining IC Layout Rules for Cost Minimization.
IEEE Journal of Solid-State Circuits SC-16(1):35-43, February, 1981.

[Saito 82] K. Saito and E. Arai.
Experimental Analysis and New Modeling of MOS LSI Yield Associated with the Number of Elements.
IEEE Journal of Solid-State Circuits SC-17(1):28-33, February, 1982.

[Sakurai 84] T. Sakurai, J. Matsunaga, M. Isobe, T. Ohtani, K. Sawada, A. Aono, H. Nozawa, T. Iizuka, and S. Kohyama.
A Low Power 46ns 256 kbit CMOS Static RAM with Dynamic Double Word Line.
IEEE Journal of Solid-State Circuits SC-19(5):578-585, October, 1984.

[Scheffer 81] L. Scheffer.
A Methodology for Improved Verification of VLSI Designs Without Loss of Area.
In C. Seitz (editor), *Proceedings of the Second Caltech Conference on VLSI*, pages 299-309. California Institute of Technology, Pasadena, CA, January, 1981.

[Schuster 78] S. Schuster.
Multiple Word/Bit Line Redundancy for Semiconductor Memories.
IEEE Journal of Solid-State Circuits SC-13(5):698-703, October, 1978.

[Seeds 67a] R. Seeds.
Yield, Economic, and Logistic Models for Complex Digital Arrays.
In *IEEE International Convention Record*, pages 60-61. March, 1967.

[Seeds 67b] R. Seeds.
Yield and Cost Analysis of Bipolar LSI.
In *IEEE International Electron Devices Meeting*, pages 12. October, 1967.

[Seth 84] S. Seth and V. Agrawal.
Characterizing the LSI Yield Equation from Wafer Test Data.
IEEE Transactions on Computer-Aided Design of Integrated Circuits and Systems CAD-3(2):123-126, April, 1984.

[Shen 85] J. P. Shen, W. Maly and F. J. Ferguson.
Inductive Fault Analysis of MOS Integrated Circuits.
IEEE Design and Test of Computers 2(6):13-26, December, 1985.

[Smith 81a] R. Smith.
Using a Laser Beam to Substitute Good Cells for Bad.
Electronics 54(15):131-134, July 28, 1981.

[Smith 81b] R. Smith, J. Chlipala, J. Bindels, R. Nelson, F. Fischer, and T. Mantz.
Laser Programmable Redundancy and Yield Improvement in a 64K DRAM.
IEEE Journal of Solid-State Circuits SC-17(5):506-513, October, 1981.

[Smith 82] R. Smith, B. Bateman, O. Sharp, P. Dishaw, and J. Smudski.
32K and 16K Static MOS RAMs using Laser Redundancy Techniques.
In *IEEE International Solid-State Circuits Conference Digest of Technical Papers*, pages 252-253. February, 1982.

[Spanos 83] C. Spanos and S. W. Director.
PROMETHEUS: A Program for VLSI Process Parameter Extraction.
In *IEEE International Conference on Computer-Aided Design Digest of Technical Papers*, pages 176-177. IEEE, September, 1983.

[Spanos 85a] C. Spanos.
Statistical Parameter Extraction for IC Process Characterization.
PhD Thesis CMUCAD-85-46, Carnegie-Mellon University, Electrical and Computer Engineering Department, May, 1985.

[Spanos 85b] C. Spanos and S. W. Director.
Statistical Parameter Extraction for IC Process Characterization.
Technical Report CMUCAD-85-58, Carnegie-Mellon University, Electrical and Computer Engineering Department, July, 1985.

[Stapper 73] C. Stapper.
Defect Density Distribution for LSI Yield Calculations.
IEEE Transactions on Electron Devices ED-20(7):655-657, July, 1973.

[Stapper 75] C. Stapper.
On a Composite Model to the IC Yield Problem.
IEEE Journal of Solid-State Circuits SC-10(6):537-539, December, 1975.

[Stapper 76] C. Stapper.
LSI Yield Modeling and Process Monitoring.
IBM Journal of Research and Development 20(3):228-234, May, 1976.

[Stapper 80] C. Stapper, A. McLaren, and M. Dreckmann.
Yield Model for Productivity Optimization of VLSI Memory Chips with Redundancy and Partially Good Product.
IBM Journal of Research and Development 24(3):398-409, May, 1980.

[Stapper 81] C. Stapper.
Comments on "Some Considerations in the Formulation of IC Yield Statistics".
Solid-State Electronics 24:127-132, February, 1981.

[Stapper 82a] C. Stapper and R. Rosner.
A Simple Method for Modeling VLSI Yields.
Solid-State Electronics 25(6):487-489, June, 1982.

[Stapper 82b] C. Stapper, P. Castrucci, R. Maeder, W. Rowe, and R. Verhelst.
Evolution and Accomplishments of VLSI Yield Management at IBM.
IBM Journal of Research and Development 26(5):532-545, September, 1982.

[Stapper 82c] C. Stapper.
Yield Model for 256K RAMs and Beyond.
In *IEEE International Solid-State Circuits Conference Digest of Technical Papers*, pages 12-13. February, 1982.

[Stapper 83a] C. Stapper, F. Armstrong, and K. Saji.
Integrated Circuit Yield Statistics.
Proceedings of the IEEE 71(4):453-470, April, 1983.

[Stapper 83b] C. H. Stapper, Jr.
Modeling of Integrated Circuit Defect Sensitivities.
IBM Journal of Research and Development 27(6):549-557, November, 1983.

[Stapper 83c] C. Stapper.
Modeling Redundancy in 64K to 16Mb DRAMs.
In *IEEE International Solid-State Circuits Conference Digest of Technical Papers*, pages 86-87. February, 1983.

[Stapper 84a] C. Stapper.
Modeling of Defects in Integrated Circuit Photolithographic Patterns.
IBM Journal of Research and Development 28(4):461-475, July, 1984.

[Stapper 84b] C. Stapper.
Yield Model for Fault Clusters Within Integrated Circuits.
IBM Journal of Research and Development 28(5):636-639, September, 1984.

[Stapper 85] C. Stapper.
The Effects of Wafer to Wafer Defect Density Variations on Integrated Circuit Defect and Fault Distributions.
IBM Journal of Research and Development 29(1):87-97, January, 1985.

[Stapper 86] C. Stapper.
Yield Statistics for Large Area ICs.
In *IEEE International Solid-State Circuits Conference Digest of Technical Papers*, pages 168-169. February, 1986.

[Strojwas 82] A. J. Strojwas.
Pattern Recognition Based Methods for IC Failure Analysis.
PhD thesis, Carnegie Mellon University, Electrical and Computer Engineering Department, October, 1982.

[Strojwas 85] A. J. Strojwas and S. W. Director.
A Pattern Recognition Based Method for IC Failure Analysis.
IEEE Transactions on Computer-Aided Design of Integrated Circuits and Systems. CAD-4(1):76-92, January, 1985.

[Sud 81] R. Sud and K. Hardee.
Designing Static RAMs for Yield as Well as Speed.
Electronics 54(15):121-126, July 28, 1981.

[Sutherland 78] I. Sutherland.
The Polygon Package.
Silicon Structures Project 1438, California Institute of Technology, Computer Science Department, February, 1978.

[Sze 83] S. Sze.
VLSI Technology.
McGraw-Hill, New York, 1983.

[Tammaru 67] E. Tammaru and J. Angell.
Redundancy for LSI Yield Enhancement.
IEEE Journal of Solid-State Circuits SC-2(4):172-182, December, 1967.

[Tarolli 81] G. Tarolli.
Private Communication.
August, 1981.

[Tarolli 83] G. Tarolli and W. Herman.
Hierarchical Circuit Extraction with Detailed Parasitic Capacitance.
In *ACM IEEE 20th Design Automation Conference Proceedings*, pages 337-345. IEEE, June, 1983.

[Taylor 84] G. Taylor and J. Ousterhout.
Magic's Incremental Design-Rule Checker.
In *ACM IEEE 21st Design Automation Conference Proceedings*, pages 160-165. June, 1984.

[Tucker 82] M. Tucker and L. Scheffer.
A Constrained Design Methodology for VLSI.
VLSI Design 3(3):60-65, May/June, 1982.

[Turley 74] A. Turley and D. Herman.
LSI Yield Projections Based Upon Test Pattern Results: An Application to Multilevel Metal Structures.
IEEE Transactions on Parts, Hybrids, and Packaging PHP-10(4):230-234, December, 1974.

[Uchida 82] Y. Uchida, T. Iizuka, J. Matsunaga, M. Isobe, S. Konishi, M. Sekine, T. Ohtani, and S. Kohyama.
A Low Power Resistive Load 64 kbit CMOS RAM.
IEEE Journal of Solid-State Circuits SC-17(5):804-809, October, 1982.

[Ueoka 84] Y. Ueoka, C. Minagawa, M. Oka, and A. Ishimoto.
A Defect-Tolerant Design for Full-Wafer Memory LSI.
IEEE Journal of Solid-State Circuits SC-19(3):319-324, June, 1984.

[Wagner 85] T. Wagner.
Hierarchical Layout Verification.
IEEE Design and Test of Computers 2(1):31-37, February, 1985.

[Walker 82] H. Walker.
Yield Simulation for Integrated Circuits - A Thesis Proposal.
VLSI Document V121, Carnegie Mellon University, Computer Science Department, October, 1982.

[Walker 83] H. Walker and S. W. Director.
Yield Simulation for Integrated Circuits.
In *IEEE International Conference on Computer-Aided Design Digest of Technical Papers*, pages 256-257. September, 1983.

[Walker 85] H. Walker and S. W. Director.
VLASIC: A Yield Simulator for Integrated Circuits.
In *IEEE International Conference on Computer-Aided Design Digest of Technical Papers*, pages 318-320. November, 1985.

[Walker 86a] D. M. H. Walker.
Yield Simulation for Integrated Circuits.
PhD thesis, Carnegie Mellon University, Computer Science Department, July, 1986.

[Walker 86b] H. Walker and S. W. Director.
VLASIC: A Catastrophic Fault Yield Simulator for Integrated Circuits.
IEEE Transactions on Computer-Aided Design of Integrated Circuits and Systems CAD-5(4):541-556, October, 1986.

[Wallmark 60] J. Wallmark.
Design Considerations for Integrated Electronic Devices.
Proceedings of the IRE 48(3):293-300, March, 1960.

[Warner 74] R. M. Warner, Jr.
Applying a Composite Model to the IC Yield Problem.
IEEE Journal of Solid-State Circuits SC-9(3):86-95, June, 1974.

[Warner 81] R. Warner.
A Note on IC-Yield Statistics.
Solid-State Electronics 24(11):1045-1047, December, 1981.

[Whitney 81] T. Whitney.
A Hierarchical Design-Rule Checking Algorithm.
Lambda (now VLSI Design) 2(1):40-43, First Quarter, 1981.

[Williams 81] R. Williams and M. Beguwala.
Reliability Concerns for Small Geometry MOSFETs.
Solid State Technology 24(3):65-71, March, 1981.

[Yamada 84] J. Yamada, T. Mano, J. Inoue, S. Nakajima, and T. Matsuda.
A Submicron 1 Mbit Dynamic RAM with a 4-Bit-at-a-Time Built-In ECC Circuit.
IEEE Journal of Solid-State Circuits SC-19(5):627-633, October, 1984.

[Yanagawa 69] T. Yanagawa.
Influence of Epitaxial Mounds on the Yield of Integrated Circuits.
Proceedings of the IEEE 57(9):1621-1628, September, 1969.

[Yanagawa 72] T. Yanagawa.
Yield Degradation of Integrated Circuits Due to Spot Defects.
IEEE Transactions on Electron Devices ED-19(2):190-197, February, 1972.

[York 85] G. York.
Triple Modular Redundancy for Yield and Reliability Enhancement in Integrated Circuits.
PhD thesis, Carnegie Mellon University, Electrical and Computer Engineering Department, October, 1985.

[Yoshida 83] M. Yoshida, M. Higuchi, K. Miyasaka, K. Shirai, and I. Tanaka.
A 288K CMOS EPROM with Redundancy.
IEEE Journal of Solid-State Circuits SC-18(5):544-550, October, 1983.

Index

ACE 51, 53, 88, 180
Active area 11
Active regions 19
Analytic yield formulas 9

Beta distribution 44, 154
Binomial distribution 17
Binomial statistics 11
Bipolar 19, 44, 173
Blocked via 23, 26, 32
Boltzmann statistics 10
Bose-Einstein statistics 15
Branch 25
Branch reconnection 66
Branch traversal 66, 83
Buried contact 66

Capacitor 24
Catastrophic defect 1
Cell failure 139
Channel net 29
Chip layout details 9
Circuit extraction 51
Circuit fault 1
Circuit fault types 56
Circuit reconnection 83
Cluster-limited yield 154
Clustering coefficient 42
CMOS 19, 173
Conclusions 173
Confidence intervals 118, 120
Contact chain 149
Converted device 25, 75
Critical area 11, 14, 15
Critical area analysis 182
Crystal defect 22
Current research 173, 175

Data contamination 167
DC topology change 23
Defect angular distribution 45
Defect category 170
Defect cluster size 44
Defect clustering 12, 41
Defect combination 170
Defect independence 36, 37
Defect label 53
Defect masking 60, 63, 71, 73, 79
Defect model approximations 35, 180
Defect model assumptions 31
Defect models 5, 19, 174
Defect neighborhood 51, 57, 92
Defect octagon 35, 53, 95
Defect probability density function 12
Defect radial distribution 12, 45, 158
Defect random number generators 3, 91, 125
Defect shape 59
Defect size distribution 4, 37, 157
Defect size distribution monitor 154
Defect spatial distribution 4, 37, 41
Defect statistics 5, 37
Defect statistics sensitivity 127, 129
Defect types 21
Diffusion pipe 22, 44, 173
Double metal 20
DRAM example 105
Dummy polygon 65

Economical process monitoring 168
Edge cracking 101
Erlang PDF 15
Excluded circuit faults 32
Excluded defect types 31
Exponential PDF 14
Extra gate material defect 75
Extra material defect 21
Extra material defect model 22

Fabrication data 6, 149, 156
Fabrication data problems 163
FABRICS II 1
Failure analysis 136
Failure analysis heuristics 137, 144
Fault analysis 3, 5, 51, 92, 180
Fault analysis procedures 57
Fault analysis sensitivity 129
Fault combination 3, 5, 59, 75, 76, 83, 92
Fault distribution 16
Fault filtering 3, 5, 59, 84, 92
Fault group 93
Fault probability 186
Fault summary 5, 94
Fault table 54
Finite resolution 20
Floating net 53
Floating transistor terminals 34

Gamma distribution 16, 42, 91
Gate oxide pinhole 29
Generated net number 80
Geometrical abstraction 19
Global defect 1, 187
Global net 36
Global process disturbance 1, 170
Graph isomorphism 54
Gross failure 44

Halting criteria 123
High-yield statistics 11

Importance sampling 183
Incomplete via 23
Incomplete via short 23
Insufficient data 164
Interdigitated comb 149
Interdigitated meander 150
Intermediate oxide pinhole 29
Intralayer short 23

Junction leakage defect 21, 30, 71
Junction leakage defect model 30

Layer combination fault analysis 54
Layer combinations 51, 89
Layout bin boundary 61
Layout bins 51, 57, 89, 178
Layout design rules 53
Layout preprocessing 51, 88
Layout simulation ix, 2
Layout tiles 54
Lithography 20
Lithography defect 22
Local circuit extraction 52
Local defect 1, 187
Local process disturbance 1

Magic 51, 88
Mask defect 159
Masking failure 171
Metal hillock 21, 29
Minimum spacing 20
Minimum width 20
Missing active material defect 71
Missing data 157, 165
Missing gate material defect 70
Missing material defect 21, 26
Missing material defect model 27
Model fitting 157
Modified Poisson statistics 12
Monte Carlo 2
Moore's half slice 15, 44
Multi-terminal device 25, 75
Murphy yield 12, 44

Negative binomial distribution 42, 45,
Negative defect 90
Net branch 25
Net fanout 62
Net loops 66
Net resistance 24
Net stitching 68
New active device fault 57, 79
New active material defect 79
New device fault 22, 24
New gate device fault 57, 75
New via fault 30, 57, 59, 71
Neyman statistics 45
NMOS 19, 26, 32
Normal distribution 91
Normal PDF 14

One-net short circuit 60
Open circuit fault 22, 25, 27, 57
Open circuit fault analysis 61

Open circuit fault analysis 61
Open device fault 27, 57, 71
Open line 26
Open via 26
Open via detection 65
Overlapped monitors 151
Oxide pinhole defect 21, 29, 71
Oxide pinhole defect model 29

Parallel plate capacitor 149
Parametric defect 1, 167
PDF 12
Photo defect 22
Photoresist pinhole 14
Poisson distribution 91
Poisson statistics 10, 42
Polygon contour count 63
Polygon decomposition 65
Polygon intersection 60
Polygon merging 58, 89
Polygon operations 58, 96, 102
Polygon package 35, 94, 179
Polygon package performance 103
Polygon representation 95
Polygon span 63
Positive defect 90
Post-processor 3
Process description 19
Process diagnosis 168
Process monitoring 149, 168
Process tables 57, 90

Radial defect distribution 46
RAPP 131, 135
RAPP examples 141, 144
Rayleigh distribution 40
Realistic process conditions 162
Rectangular PDF 12
Redundancy analysis post-processor 3, 6, 131
Reference area 14
Repeating defect 159
Research motivation v

Serpentine meander 149
Short circuit fault 22, 57
Short circuit fault analysis 60
Shorted device fault 27, 57, 73
Single-wafer yield 17
Spare swapping procedure 140
SRAM example 109
Step coverage breakage 21
Substrate net 30
Susceptible area 11

TAPP 170
Test analysis post-processor 170
Test category 170
Transistor gate 25
Transistor size change 24
Transistor terminal 25
Triangular PDF 12
Truncated defect size distribution 41, 91, 116
Type-driven fault analysis 56

Unblocked via 33
Uniform distribution 41, 91

Virtual artwork 185
VLASIC 2
VLASIC accuracy 117
VLASIC applications 3
VLASIC control loop 92
VLASIC examples 105
VLASIC fast implementation 175
VLASIC hierarchical implementation 177
VLASIC implementation 87
VLASIC parallel implementation 176
VLASIC performance 111
VLASIC sensitivity analysis 127
VLASIC space usage 178
VLASIC system structure 3
VLASIC tuning 6
Vulnerable area 11

Wafer map 89
Wafer-scale integration 45
Winding number 71, 99

Yield bounds 17
Yield modeling 9, 132
Yield simulation v, 2, 134